George Henslow

The Origin of Floral Structures through Insect and other Agencies

George Henslow

The Origin of Floral Structures through Insect and other Agencies

ISBN/EAN: 9783337272630

Printed in Europe, USA, Canada, Australia, Japan

Cover: Foto ©berggeist007 / pixelio.de

More available books at **www.hansebooks.com**

THE INTERNATIONAL SCIENTIFIC SERIES

THE ORIGIN OF FLORAL STRUCTURES

THROUGH INSECT AND OTHER AGENCIES

BY THE

REV. GEORGE HENSLOW, M. A., F. L. S., F. G. S.

PROFESSOR OF BOTANY, QUEEN'S COLLEGE, AND
LECTURER TO ST. BARTHOLOMEW'S HOSPITAL MEDICAL SCHOOL, LONDON
AUTHOR OF "EVOLUTION AND RELIGION," "CHRISTIAN BELIEFS RECONSIDERED,"
"BOTANY FOR CHILDREN," "FLORAL DISSECTIONS," ETC.

WITH EIGHTY-EIGHT ILLUSTRATIONS

NEW YORK
D. APPLETON AND COMPANY
1888

PREFACE.

The belief that we must look mainly to the environment as furnishing the influences which induce plants to vary in response to them—whereby adaptive morphological (including anatomical) structures are brought into existence—appears to be reviving. To illustrate the progress of this belief, I will give a few cases.

In 1795, Geoffroy Saint Hilaire "seems to have relied chiefly on the conditions of life, or the 'monde ambiant,' as the cause of change." *

In 1801, Lamarck "attributed something to the direct action of the physical conditions of life" as the means of modification, "something to the crossing of already existing forms, and much to use and disuse."

In 1831, Mr. Patrick Matthew (who, like Dr. W. C. Wells in 1818, anticipated Mr. Darwin in the theory of "natural selection") "seems to have attributed much influence to the direct action of the conditions of life."

* I quote from Mr. Darwin's "Historical Sketch" in his *Origin of Species*, 6th ed., 1878.

In 1844, the "Vestiges of Creation" appeared. The author suggests that "impulses" were imparted to the forms of life, on the one hand advancing them, and on the other hand tending to modify organic structures in accordance with external circumstances; the effects thus produced by the conditions of life being gradual.

In 1852, Mr. Herbert Spencer "attributed the modifications [of species] to the change of circumstances."

In 1859, "The Origin of Species" appeared. Mr. Darwin did not at first seem to lay so much stress as his predecessors upon the action of the environment as a cause, for he says: "It is curious how largely my grandfather, Dr. Erasmus Darwin, anticipated the views and erroneous grounds of opinion of Lamarck." Again, in speaking of the constancy of some varieties, he says, "Such considerations incline me to lay less weight on the direct action of the surrounding conditions, than on a tendency to vary, due to causes of which we are quite ignorant."* He had, however, previously said, "Changed conditions of life are of the highest importance in causing variability. . . . It is not probable that variability is an inherent and necessary contingent under all circumstances." †

With regard to my own opinion, having been early and greatly interested in Paley's "Natural Theology," as well as the "Vestiges" when Mr. Darwin's work

---

* *Or. of Sp.*, p. 107.   † *Ibid.*, p. 31.   See also *Desc. of Man*, ii., p. 388.

PREFACE.

appeared, the great difficulties I felt in accepting natural selection as any real *origin* of species lay, first, in the seeming impossibility of the histological minutiæ of the organs in adaptation having been selected together; and, secondly, in the idea that all those wonderful and "purposeful" structures which Paley thought could only have been "designed," could be the ultimate result of any number of accidental and apparently at first "purposeless" variations. In a broad sense natural selection seemed obviously true; for Geology had revealed the fact that the world had been peopled over and over again by old forms dying out and new forms coming in; so that although it might account for the extinction of the former, it did not seem to me capable to account for the origin of the latter. I, therefore, still looked to the environment as affording a better clue to the source of variations.*

In 1869, when watching a large humble-bee hanging on to the dependent stamens of *Epilobium angustifolium*, the idea first occurred to me that insects themselves might be the real cause of many peculiarities in the structure of flowers. The thought passed through my mind that the way the stamens hung down might perhaps have become an hereditary effect from the repeatedly applied weight of the bees.

In 1877, I advanced this idea as a speculation

* See Letter to *Nature*, vol. v., p. 123.

when suggesting the origin of nectaries and irregularities of flowers in my paper on "Self-fertilisation of Flowers." *

In 1880, Mr. A. R. Wallace reviewed Dr. Aug. Weismann's "Studies in the Theory of Descent." † In this work the author says: "According to my view, transmutation by purely internal causes is not to be entertained. . . . The action of external inciting causes is alone able to produce modifications." Mr. Wallace adds that he had "arrived at almost exactly similar conclusions."

In 1881, when reviewing Paul Janet's work on "Final Causes," ‡ I took occasion to remark that "I regarded the environment as by far the most important "cause" of variations, in that it influences the organism, which, by its inherent but latent power to vary, responds to the external stimulus, and then varies accordingly."

In 1881, appeared the first really systematic treatise that I know of, by Dr. C. Semper,§ which dealt with the origin of variations in animals as being referable to the environment.

In 1884, Dr. A. de Bary's "Comparative Anatomy of the Vegetative Organs of the Phanerogams and Ferns,"

---

* *Trans. Lin. Soc.*, 2nd ser., Bot., vol. i., p. 317.
† *Nature*, xxii., p. 141.   ‡ *Modern Review*, 1881, p. 53.
§ "*The Natural Conditions of Existence as they affect Animals,*" Intern. Sci. Ser., vol. xxxi.

was published in English. In the Introduction, the author writes as if it were a perfectly well understood thing that species have arisen by adaptations to the influences of the environment.*

In 1886, Mr. Herbert Spencer contributed two articles on "The Factors of Organic Evolution" to the *Nineteenth Century*.† In these he showed, from many passages in Mr. Darwin's works, especially "Animals and Plants under Domestication" and in his later volumes, that he became much more favourably inclined to the belief that the effects of the environment were accumulative, and that in the course of some generations the variations set up tended to cease and become fixed. Mr. Spencer particularly notes the change of view, as illustrated by the expression "little doubt" being replaced by "no doubt" in the following sentence: "I think there can be no doubt that use in our domestic animals has strengthened and enlarged certain parts, and disuse diminished them; and that such modifications are inherited." ‡ It may be added that in "The Cross and Self Fertilisation of Flowers" (1876), and in "Forms of Flowers" (1877), Mr. Darwin makes many observations upon the effects of the external conditions upon plants as influencing and modifying them in various ways. It is curious to note that the three influences upon which Lamarck laid

---

* See, *e.g.*, p. 25. † See p. 570 and p. 749.

‡ "Use" and "disuse" in animals corresponds to what I have called "hypertrophy" and "atrophy" in plants, in this work.

emphasis are just those which Mr. Darwin himself latterly, though often indirectly perhaps, laid stress upon in his experiments, viz. crossing, use and disuse, and the physical conditions of life.

In 1886, also appeared an article in *Nature*, entitled, " Plants considered in Relation to their Environment." It was not signed, but the author alludes to the external conditions as bringing about all sorts of changes in the vegetative system. He stops short of discussing floral structures.

In 1886, Dr. Vines' " Physiology of Plants " appeared. After discussing various views and theories of reproduction, he observes, that " variability was first induced as the response of the organism to changes in the conditions of life." * . . . We conclude, then, that the production of varieties is the result of the influence of the conditions of life. †

In the last page of his work, Dr. Vines calls attention to Naegeli's view as follows : " Naegeli suggests, and his suggestion is worthy of serious consideration, that there is an inherent tendency to a higher organisation, so that each succeeding generation represents an advance, . . . as in cases of what is termed *saltatory* evolution." Thus,

---

* Page 676. Dr. Vines here uses almost identically the same words as myself in 1881. I have just found that Mr. St. G. Mivart said much the same in 1870, *Genesis of Species*, p. 269. See also O. Schmidt's *Doctrine of Descent and Darwinism*, p. 175.

† Page 679.

while Mr. Darwin seems at last to have tacitly accepted Lamarck's ideas, at least to a considerable extent, we have here a return in 1887 to the views of the author of the "Vestiges" of 1884.

1888. I have attempted in the present work to return to 1795, and to revive the "Monde ambiant" of Geoffroy Saint Hilaire, as the primal cause of change. My object is to endeavour to refer every part of the structures of flowers to some one or more definite causes arising from the environment taken in its widest sense. To some extent the attempt must be regarded as speculative; and, therefore, any deductive or *à priori* reasonings met with must be considered by the reader as being suggestive only.

CONTENTS.

CHAPTER		PAGE
I.	GENERAL PRINCIPLES	1
II.	THE PRINCIPLE OF NUMBER	7
III.	THE PRINCIPLE OF NUMBER—*Continued*	25
IV.	THE PRINCIPLE OF ARRANGEMENT	39
V.	THE PRINCIPLE OF COHESION	48
VI.	THE PRINCIPLE OF COHESION—*Continued*	54
VII.	THE PRINCIPLE OF COHESION—*Continued*	62
VIII.	THE PRINCIPLE OF ADHESION	78
IX.	THE CAUSE OF UNIONS	84
X.	THE RECEPTACULAR TUBE	89
XI.	THE FORMS OF FLORAL ORGANS	101
XII.	THE ORIGIN OF "ZYGOMORPHISM"	116
XIII.	THE EFFECTS OF STRAINS ON STRUCTURES	123
XIV.	ACQUIRED REGULARITY AND "PELORIA"	128
XV.	THE ORIGIN OF FLORAL APPENDAGES	133
XVI.	SECRETIVE TISSUES	140
XVII.	SENSITIVENESS AND IRRITABILITY OF PLANT ORGANS	151
XVIII.	ORIGIN OF CONDUCTING TISSUES	164

CHAPTER		PAGE
XIX.	Colours of Flowers	174
XX.	The Emergence of the Floral Whorls	184
XXI.	The Development of the Floral Whorls	191
XXII.	Heterogamy and Autogamy	198
XXIII.	Heterostylism	203
XXIV.	Partial Diclinism	220
XXV.	Sexuality and the Environment	230
XXVI.	Degeneracy of Flowers	251
XXVII.	Degeneracy of Flowers—*Continued*	273
XXVIII.	Progressive Metamorphoses	285
XXIX.	Retrogressive Metamorphoses	295
XXX.	Phyllody of the Floral Whorls	301
XXXI.	The Varieties of Fertilisation	311
XXXII.	Fertilisation and the Origin of Species	329

LIST OF ILLUSTRATIONS.

FIGURE		PAGE
1.	Diagram of a typical flower composed of six whorls ...	3
2.	Diagram of the positions of opposite leaves, illustrating the method of passage to alternate arrangements	11
3.	Diagrams of floral æstivations, showing the passage from the two-fifth or quincuncial, to the contorted	15
4.	Diagram of flower of *Garidella*, with stamens superposed to petals	21
5.	Diagram of flower of *Helleborus niger* with stamens superposed to twenty-one nectaries	22
6.	Diagrams illustrating the anatomy of the floral receptacle of a Wallflower, showing the origin of the floral members ...	32
7.	Diagram of the leaf-traces in the stem of *Arabis albida* ...	39
8.	Vertical and transverse sections of the wall of the inferior ovary of *Campanula medium*, showing how the sepaline cords originate those of the rest of the floral organs (see fig. 15, p. 71)	43
9.	Flower of *Phyteuma*, showing cohesion by contact and congenital, in the corolla	50
10.	Flower of *Mimulus* undergoing "dialysis"	51
11.	Stamens of *Centaurea*, showing syngenesious anthers; method of fertilisation by "piston-action" (*b*), nectary and direction of insect-proboscis, etc.	60
12.	Anatomy of the floral receptacle of Hellebore, showing the changes in orientation of the cords	64
13.	Anatomy of the floral receptacle of *Pelargonium*, showing changes in the orientation, in the separation and in the union of the cords	65

xvi LIST OF ILLUSTRATIONS.

FIGURE | PAGE
14. Anatomy of the floral receptacle of Ivy, showing the multiplication and differentiation of the cords, etc. 68
15. Anatomy of the floral receptacle of *Campanula medium*, showing the distribution of the cords, etc. (see fig. 8, p. 43) 71
16. Origin and development of the ovule in *Beta* 73
17. Carpels of *Acer*, showing the thickened bases, preparatory for ovules 75
18. A separate carpel of *Primula sinensis*, with marginal ovules and a "heel-like" process, the origin of the free central placenta 76
19. Anatomy of the floral receptacles of *Lysimachia* and *Primula*, showing the cords of five carpels 77
20. *Echium*, showing declinate stamens and protandrous condition 82
21. Ovary, stamens, and stigmas of *Aristolochia* 83
22. Vertical sections of buds of *Pyrus* and *Cotoneaster*, showing degrees of adhesion or undifferentiated condition between the ovary and receptacular tube 90
23. *Orchis Morio* (?), with arrest of pistil, the receptacular tube represented by a rod-like pedicel. Two anthers are developed instead of one (a) 92
24. Receptacular tube of Rose, bearing a leaf and a stipular sepal 93
25. Vertical section of the receptacular tube of Hawthorn, with supernumerary carpels arising from the summit 93
26. Leaves of Pear with hypertrophied and sub-fasciate petioles 94
27. *Fuchsia* with foliaceous sepals, partly detached from the ovary 94
28. Anatomy of the receptacular tube of *Prunus*, showing the origin of the petaline and staminal cords 95
29. Part of the receptacular tube of Cherry, showing the distribution of cords in the sepaline lobes 97
30. Anatomy of inferior ovary of *Alstrœmeria*, showing the junction between the ovary and the tube 97
31. Flower of *Duvernoia*, showing its adaptability for intercrossing 107
32. Flower of *Calceolaria*, showing thickened ridges, etc., and adaptability for intercrossing 109
33. Flower of *Dictamnus*, showing declinate stamens and displacement of petals 110
34. Flower of *Epilobium angustifolium*, showing dependent stamens and displacement of petals 111

LIST OF ILLUSTRATIONS.

FIGURE	PAGE
35. Flower of *Veronica Chamædrys*, showing method of fertilisation by insects, and degeneracy of anterior petal	111
36. Flower of *Teucrium*, to show effect of weight of insect with exposure of stamens	117
37. Diagrams of *Narcissus cernuus*, to show instability in the heterostylism and lengths of stamens	121
38. Basal end of a Pear, to show cause of thickening in response to forces	124
39. Diagram of a declinate bough, showing distribution of forces	125
40 a. A diagram of declinate stamens, to show distribution of forces	126
40 b. Flower of *Lamium album*, to show distribution of forces ...	126
41. Base of flower of *Amaryllis*, showing the honey-protector ...	134
42. Adhesive epidermal cells of roots of Orchids	137
43. Stipules of *Impatiens*, showing nectariferous tissue	140
44. Petals passing into nectariferous stamens of *Atragene* ...	141
45. Cells of hair of *Tradescantia*, showing the state of protoplasm before and after excitation by electricity	152
46. Climbing peduncle of *Uncaria*, thickened after irritation by the support	156
47. Flower of *Genista tinctoria*, before and after mechanical irritation; the claws of the keel and wing petals being in unstable equilibrium	160
48. Flowers of *Lopezia* in three stages, showing movements of the staminode and stamen	161
49. Flower of *Medicago sativa*, before and after mechanical irritation, the staminal tube being in unstable equilibrium ...	162
50. Transverse sections of conducting tissues of *Fumaria*, *Rubus*, and of a Crucifer	164
51. Diagram of emergence of the petaline stamens of *Peganum* outside the sepaline	189
52. Flower-bud, and same opened, of *Stellaria media*, showing conditions of degeneracy and adaptations for self-fertilisation	255
53. Flower-bud, and essential organs of *Epilobium montanum*, showing positions for self-fertilisation	255
54. Styles and stigmas of the two forms of Pansy, showing the conditions which (a) prevent and (b) secure self-fertilisation, respectively	255

LIST OF ILLUSTRATIONS.

FIGURE	PAGE
55. Styles and stigmas of self-fertilising forms of Pansy	257
56. Details of structure of cleistogamous Violets	258
57. Details of structure of cleistogamous *Oxalis Acetosella*	260
58. Flower-bud and stamens of cleistogamous *Impatiens*	261
59. Flower-bud and section of cleistogamous *Lamium amplexicaule*	261
60. Corolla, stamens, and style of *Salvia clandestina*, showing adaptations for self-fertilisation	262
61. Transitional forms between bracts and leaves of *Helleborus viridis*	286
62. Inflorescence of *Cornus florida*, showing floral mimicry	287
63. Inflorescence of *Darwinia*, showing floral mimicry	287
64. Involucral bract of *Nigella*, bearing an anther	288
65. Glumes of *Lolium*, both antheriferous and stigmatiferous	288
66. *Ranunculus* with a foliaceous sepal	289
67. Foliaceous calyx of *Trifolium repens* with stipulate leaves, borne by the receptacular tube	289
68. Flower and leaf of *Mussœnda*	290
69. *Linaria* with one sepal petaloid	291
70. Calyx of Garden Pea with carpellary lobes	292
71. Ovuliferous sepal of Violet	292
72. Corolla of Foxglove with filamentous processes, some being antheriferous	292
73. *Aquilegia*, the corolla with polleniferous spurs	293
74. Ovuliferous petals, etc., of *Begonia*	293
75. Ovuliferous anthers of *Sempervivum*	294
76. Stigmatiferous and ovuliferous stamens of *Begonia*	294
77. Carpels and ovules originating from a placenta of Carnation, the carpels again ovuliferous (*a*)	295
78. Stameniferous carpels of Willow and *Ranunculus auricomus*	296
79. Petaliferous placentas of *Cardamine pratensis* and of *Rhododendron*	296
80. Metamorphosed sub-petaloid carpel of *Polyanthus*	297
81. Foliaceous connective of *Petunia*	298
82. Petalody, or "hose-in-hose" form, of connective in a double Columbine (*Aquilegia*)	298
83. Foliaceous stamen and petal of the Alpine Strawberry and stamen of the Green Rose	302

FIGURE	PAGE
84. Stamen of *Jatropha Pohliana*, with foliaceous membranes to the anther-cells	302
85. Metamorphosed and foliaceous ovules of Mignonette	305
86. Metamorphosed and foliaceous ovules of *Sisymbrium Alliaria*	306
87. Tubular excrescence on the labellum of Cattleya, homologous with an ovule	306
88. Multifold carpels, with ovuliferous margins, from a malformed Primrose	308

THE ORIGIN OF FLORAL STRUCTURES
THROUGH INSECT AND OTHER AGENCIES.

CHAPTER I.

GENERAL PRINCIPLES.

INTRODUCTORY.—Much has been written on the structure of flowers, and it might seem almost superfluous to attempt to say anything more on the subject; but it is only within the last few years that a new literature has sprung up, in which the authors have described their observations and given their interpretations of the uses of floral mechanisms, more especially in connection with the processes of fertilisation.

Moreover, there is a considerable amount of scattered literature on special points which seems never to have been collated, so as to show the reletive significance of the different classes of observations to which the authors have devoted themselves respectively. The consequence is, that, good as each in itself may be, it often requires the help of other classes of facts to enable one to fully elucidate any question to be discussed.

Now, the primary object of the first really scientific study of plants was their classification, and no longer with the sole view of ascertaining the real or imaginary medicinal uses of herbs; as had been the case in Gerarde's time, when a botanist and a herbalist were one and the same.

Systematic botanists, however, have hitherto invariably contented themselves with observing differences of structure only; and paid little or no attention to the "why" and the "wherefore" of the differences they seized upon as being more or less important for the purpose of distinguishing species. When, however, the desirability of a more thorough knowledge of the origin of parts of plants as interpreting morphological characters was felt, developmental history began to be studied; a method strongly insisted upon by Schleiden, for example; and the most elaborate result of this method of investigation is undoubtedly Payer's *Traité d'Organogénie Comparée de la Fleur*, published in 1857: but if it be thought sufficient to limit the study of flowers to tracing their morphological development alone, one soon begins to see that it is far from being so, and, taken by itself, it may lead one into false interpretations, so that to the study of development must be added that of anatomy. To Ph. van Tieghem we are indebted for an elaborate treatise, entitled *Recherches sur la Structure du Pistil et sur l'Anatomée Comparée de la Fleur* (1871), dealing with the more minute details of floral structures. This treatise, however, still leaves much to be desired.

Besides these methods, analogy and especially teratology furnish assistance of no mean value. Here we are especially indebted to Dr. M. T. Masters for his standard work on *Teratology*.*

Now, any one of these methods taken alone would be insufficient, and in many cases would be far from thoroughly accounting for particular points under consideration.

Hence to arrive at a complete interpretation of the origin of every sort of structure to be found in flowers, it can only

* A German edition, *Pflanzen Teratologie*, ed. Dammer, 1886, has numerous additions.

be done by calling in the aid of each and all these methods to the very utmost extent possible.

Lastly, to attempt any theoretical exposition of the evolutionary history of flowers, considerable caution is required; for the causes of variation are generally so obscure, the chances of seeing them in activity so small, and experimental methods of verification well-nigh impossible, that speculations on this subject cannot altogether escape the bounds of hypothesis so as to become demonstrable facts. Hence observations which I shall make later on, with reference to the origin of existing floral structures, will not profess to be anything more than theoretical, and at most only a "working hypothesis" for future investigations.

THE STRUCTURE OF A TYPICAL FLOWER.—Before considering how the innumerable forms of flowers deviate from one another, it is advisable to assume some typical form or plan as a preliminary basis to start from, or to which all flowers, if possible, may be referred as a standard. It would be quite possible to adopt some kind of flower as it exists in nature, but as this would be arbitrary, it may be better to take an ideal

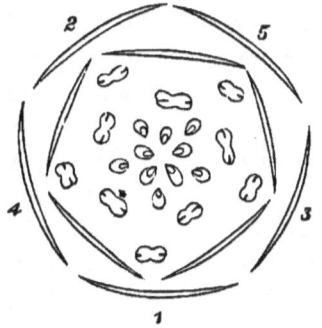

Fig. 1.—Diagram of a typical flower.

type, and the diagram (Fig. 1) will answer the purpose, in which the outermost circle is supposed to represent a cross section of the five Sepals constituting the Calyx. The second circle is that of the five Petals of the Corolla. The third stands for the Anthers of the five Stamens superposed to the sepals; the fourth being those of five Stamens superposed to the petals. These two whorls of stamens together constitute the Androecium. Lastly, there are represented two * whorls of Carpels forming the Gynœcium † or Pistil. The outermost whorl of carpels is superposed to the sepals, the innermost to the petals.

There may be additional structures in flowers, such as disks, honey-glands, etc.; but as these, when they occur on the floral-receptacle, are merely cellular protuberances and form no part of the floral whorls proper—not being *foliar* in their origin—they may be omitted, especially as their position is by no means constantly the same in all flowers.‡

THE PRINCIPLES OF VARIATION.—Having thus assumed an ideal type, we may at once consider the "Principles of Variation," as I propose to call them, in accordance with which the different members of flowers can be altered; so that by means of various combinations of these principles all the flowers in the Vegetable Kingdom can be brought under this one fundamental plan.

There are five principles which require special consideration. They are usually designated by the terms Number, Arrangement, Cohesion, Adhesion, and Form.

"Number" refers to the number of whorls and the number of parts in each whorl. If two or more whorls contain the same number of parts or be multiples of one another, they are said to be "symmetrical" or "isomerous." If they differ in the number of parts they are "unsymmetrical" or "anisomerous."

"Arrangement" refers to the relative positions of the

* Why I assume *two* whorls for the pistil, instead of *one* only, as is generally done, will be understood hereafter. I have since found that Robert Brown came to the same conclusion (*Col. Works*, i. 293).

† I adopt the spelling *Gynœcium* for the sake of uniformity; it may be regarded as a shortened form of *Gynæcœcium*.

‡ I do not here allude to certain glandular structures, which may be the homologues of arrested organs.

different whorls, as well as of those of the individual members of the whorls with regard to each other.

"Cohesion" signifies the union of parts of any, but of the same whorl. The original or ancestral condition of the parts composing every whorl is presumed, on the principles of evolution, to have been one of entire freedom; so that the members were as completely separate or free as, for example, they are in a Buttercup. Reversions to this condition of freedom may occur, and then the process is called "dialysis," as in the case of a polypetalous *Campanula* occasionally cultivated as a garden plant.

"Adhesion" signifies the union of parts of different whorls; as well as that between the ovary and the receptacular tube, constituting the so-called inferior ovary. I regard adhesion as representing a more advanced or a more highly differentiated state than that of cohesion. Reversions may occur by "solution," which brings about a freedom of parts normally united, as in the abnormal cases of Apples, double Saxifrage, members of the *Umbelliferæ*, etc., which have all their parts perfectly free, though with inferior ovaries under ordinary circumstances.

"Form" refers to the shape of the organs; such as those of sepals and petals upon which generic characters are so often founded, the length of the filaments, and other peculiarities. If all the parts of any whorl be exactly alike, it is said to be "regular;" if not, the whorl will be "irregular."

The above five principles constitute the most important in accordance with which Nature has brought about the infinite diversity which exists in the Floral world. There are minor distinctions, hereafter to be considered, such as colours, scents, etc.; but they are of less importance in investigating the causes at work which have evolved specific and generic differences amongst flowering plants.

There is another point which may be here noticed. That a flower-bud is a metamorphosed leaf-bud is now an accepted fact; but an obvious difference between them consists in the arrested state of the axis of the former, constituting the floral receptacle; and the question arises, how has this arrest been brought about? Like all other peculiarities of structure to be described, I would attribute the arrest primarily to the altered nature of the foliar organs on becoming members of flowers. Thus, a Fir-cone and a Buttercup are arrested branches; but when the parts of a flower are reduced in number, and instead of being in a continuous spiral are grouped in "compressed cycles,"* I would then (hypothetically) attribute this further reduction of the axis, as well as other features hereafter to be described, to the irritation of insects in probing for juices, and causing nectaries to be formed.† It is the commonest thing for leaf-buds to be arrested, and sometimes metamorphosed as well, by insects puncturing and depositing their eggs in them. Such may be seen on the terminal shoots of Yews, Thyme, and in certain kinds of Oak-galls, etc. In all such cases the immediate effect is the total arrest of the axis, though the leaves may be but slightly altered, as in the Yew. How the various metamorphoses of leaves into petals, etc., has followed will be discussed later on.

It must not be forgotten, however, that the tendency to shorten the axis is primarily, in some cases, due to the altered structure of the foliar organs, as in Gymnosperms; whereby they undertake the reproductive functions. At the same time, I think insects have had a good deal to do with it, in many other phanerogams, which have but few parts to their whorls.

Each of the above principles must now be considered in detail.

* See pp. 41, 42. † See p. 140, seqq.

CHAPTER II.

THE PRINCIPLE OF NUMBER.

NUMBER — GENERAL OBSERVATIONS.—The first principle of Variation to be considered is that of the number of parts composing the different whorls of flowers. There are good reasons for considering that six whorls, consisting of five, four, three, or two parts each, as the case may be, should be regarded as the theoretically complete number of verticils of any flower.

Anatomical investigations prove that the rule is for the pedicel to contain—at least, immediately below the flower,—if the latter be pentamerous, ten more or less distinct fibrovascular cords, five of which belong to the sepals and five to the petals; if it be hexamerous, there will be six cords, three for each whorl of the perianth. Each of these cords can give rise by branching, first, to a whorl of stamens and subsequently to a whorl of carpels, furnishing at least two marginal and one dorsal cord for each of the latter.

In many flowers both whorls of stamens are present, and the andrœcium is then isomerous with the entire perianth. More often one whorl is arrested, and then it may be either one; but most usually it is the petaline. On the other hand, the calycine may not be developed as in Primroses, *Rhamnus*, etc.

The absence of the petaline stamens is possibly attributable to the law of compensation, in consequence of the enhanced growth of the corolla, the petals thereby abstracting the nourishment that would be required by the stamens superposed to them.

That the number of staminal whorls should be two in verticillate flowers, *i.e.*, equal to the perianth, is apparent from the fact that two whorls prevail in Monocotyledons and are not at all uncommon in Dicotyledons; and when the petaline whorl alone exists, as in *Primulaceæ* and *Myrsineæ*, calycine staminodia are sometimes present which tend to restore the complete number, as in the genus *Samolus* in the former and in the tribe *Theophrasteæ* of the latter order.

The reduction of the number of carpels is very generally carried to a greater extent than that of the stamens. Assuming two complete whorls of carpels as the primitive number, not only are both rarely to be found in the same flower, as in *Butomus*, but a portion only of one whorl is commoner than even a single entire whorl. Thus, two are characteristic of *Cruciferæ*, *Polygaleæ*, and of most of the gamopetalous orders; while one carpel only prevails in *Leguminosæ* and elsewhere.

That the absence of parts of, as well as of entire whorls of flowers as they now exist does not represent primitive conditions, is testified to by the frequent occurrence of various kinds of degradations, such as were alluded to above in the case of the staminodia of *Samolus*, etc. Thus, with regard to the calyx, it is a noticeable fact that when the inflorescence consists of a large number of flowers, especially if small and closely compacted, there is a strong tendency for the sepals to become partially arrested and remain rudimentary, or even not to be developed at all. This is particularly observable in some epigynous orders as *Umbelliferæ*, *Araliaceæ*, *Caprifoliaceæ*, *Rubiaceæ*, *Compositæ*, etc.

THE PRINCIPLE OF NUMBER. 9

The degradation of the corolla is likewise very common. As its enhancement has been due to insect agency, so, conversely, its reduction in size, colour, etc., is presumably often the result of the neglect of insects. Consequently inconspicuousness becomes a characteristic feature of self-fertilising flowers. By increased degradation the corolla may disappear entirely, as in *Sagina apetala*, some cleistogamous flowers, and in the *Incompletæ* generally. Such degradation is also characteristic of wind-fertilised flowers.

As both calyx and corolla may be degraded and disappear, so may the stamens and carpels, unisexual and neuter flowers being the result.

Further observations, however, will be made upon this subject when treating of the several whorls respectively, and especially when discussing the phenomenon of degeneracy.

THE ORIGIN OF DIFFERENT NUMBERS. The number of parts constituting the floral whorls is, without doubt, primarily due to phyllotaxis; and therefore, to understand why certain numbers, such as fives, fours, and threes prevail, it is needful to give some preliminary remarks on the principles of leaf arrangement. It has long been observed that these are referable to two kinds—one in which two or more leaves are situated on the same node, when they are decussate,* that is to say, each pair or whorl of three or more leaves alternates in position with the whorl immediately above and below it. The second system is when only one leaf occurs at a node; the leaves are then said to be alternate. The leaves are then arranged on a continuous spiral line, and can be represented by the fractions of the well-known series $\frac{1}{2}, \frac{1}{3}, \frac{2}{5}, \frac{3}{8}, \frac{5}{13}, \frac{8}{21}$, etc. Of these fractions the denominator represents the number of

---

* Rare exceptions occur in species of *Potamogeton*, in which alternate internodes between the distichously arranged leaves are suppressed, so that they become opposite, but are all in the same plane.

leaves in a "cycle," and the numerator the number of times a spiral line, passing through the position of the leaves, coils round the stem in forming a cycle; thus, with the $\frac{2}{5}$ arrangement, any leaf being taken as number 1, the sixth leaf will be first that falls in the same vertical line with number 1, the leaves 1 to 5 constituting the cycle. The portion of the spiral line which passes through the leaves 1 to 6 coils *twice* round the stem, and if projected on a plane forms two circles. The angular distance, measured from the centre of the stem or circles, between any two successive leaves is always found by multiplying 360° by the fraction: thus $\frac{2}{5} \times 360° = 144°$.

The interpretation, therefore, of the prevailing numbers 3 and 5 in floral whorls is that they are, in most cases, cycles of the $\frac{1}{3}$ or $\frac{2}{5}$ types respectively; while 4 is primarily due to the union of two pairs of opposite and decussate parts. 6, 8, 10 are merely the doubles of the preceding, and mostly represent two pairs of whorls or cycles blended together, thus forming one whorl, or so closely approximated as scarcely recognizable as two; though the rare number 8, in some cases, such as *Nigella*, and *Helleborus fœtidus*, may represent a cycle of the $\frac{3}{8}$ type. Similarly, the still rarer numbers 7, 9, and 11 in flowers correspond to the absence of these numbers as denominators of any fractions of the above prevailing series.

With the exception of dimerous and tetramerous whorls, all the rest are presumably due to alternate arrangements. Now, opposite leaves present a more primitive type than alternate; that this is so, is not only reasonable from the primordial condition of the cotyledons of Dicotyledons, but the transition from an opposite to an alternate condition may be often witnessed on rapidly growing stems, such as of the Jerusalem Artichoke. Whenever this plant bears opposite leaves below, and alternate leaves above, it will be found that the arrangement of the latter is almost invariably represented by the ⅖ type. It is secured by developing internodes between the two opposite leaves of each pair, and by shifting their positions so as to acquire ultimately an angular divergence of 144°.*

The feature to be especially observed in the transitions from opposite to alternate arrangements is the order in which the opposite leaves separate so as to assume successive positions on the continuous spiral line passing through their insertions, when they have become alternate. This will be understood from the accompanying diagram, in which the numbers represent the order which the leaves will ultimately assume on the ⅖ type; though they are placed as if still opposite and decussate. The numbers 1 and 2, 3 and 4, 5 and 6, etc., represent the successive pairs of opposite leaves, the arrows showing the direction of the spiral.

It will be at once observed that the numbers 6, 9, 14, and 22 are in the same row, and correspond to the divergences $\frac{2}{5}$, $\frac{3}{8}$, $\frac{5}{13}$, $\frac{8}{21}$. No. 17 falls into the series $\frac{3}{8}$, and completes the second cycle of that type from No. 1.

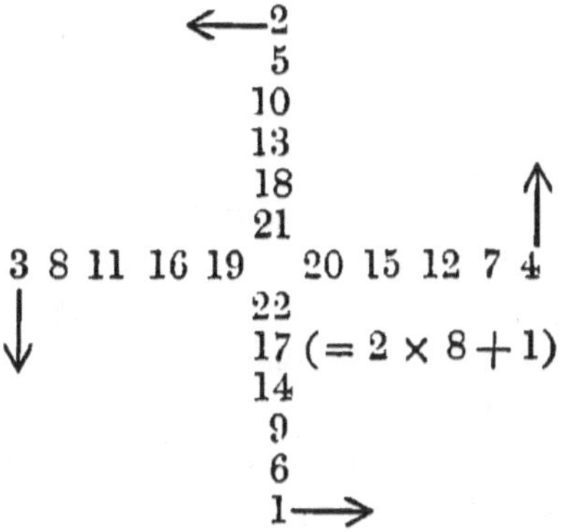

```
          ←—2
           5
          10
          13
          18
          21                    ↑
3 8 11 16 19    20 15 12 7 4
          22
          17 (= 2 × 8 + 1)
          14
           9
           6
           1—→
```

Fig. 2.—Opposite leaves passing into alternate.

It may be observed here, as occasion will arise for a fuller allusion to the significance of the fact, that, with the sole exception of the

---

* I have fully explained this in my paper, *On the Variations of the Angular Divergences of the Leaves of* Helianthus Tuberosus, Trans. Lin. Soc., vol. xxvi., p. 647. See also *On the Origin of the Prevailing Systems of Phyllotaxis*, l.c., 2nd series, vol. i. p. 37.

distichous or $\frac{1}{2}$ type, every other arrangement always has *three leaves in every projected circle.*

It may be noticed that No. 4 not only does not occur in the row 1, 6, 9, etc.,—a fact which corresponds with the rarity of a ternary arrangement occurring amongst flowers of Dicotyledons,—but in order to fall over No. 1 it would have to pass through 270°, that is from right to left, practically an impossibility; so that when "threes" are met with in Dicotyledons we must look for some other interpretation than to refer them to the $\frac{1}{3}$ type.

The numbers 7 and 11, as stated, are extremely rare in flowers, and this is in accordance with the fact that they belong to another series, viz. $\frac{1}{3}$, $\frac{1}{4}$, $\frac{2}{7}$, $\frac{3}{11}$, $\frac{5}{18}$, etc , which is rarely represented in nature. Examples, however, will be found in the leaves of *Sedum reflexum*, on some branches of *Araucaria imbricata*, and sometimes in the Jerusalem Artichoke. In the last case, it will be discovered that the heptastichous or $\frac{2}{7}$ type arises out of verticils of threes, in precisely the same way as the pentastichous or $\frac{2}{5}$ type does from an opposite and decussate arrangement; and as there are always *four leaves in every projected circle*, for every type of this series, excepting the first or $\frac{1}{3}$, it can only occur where the leaves are narrow or are short, or do not occupy too much space so as to overshadow one another.

VARIATIONS IN THE FLORAL SYMMETRY.—Besides the fact that certain numbers are often characteristic of certain species, genera, or even orders, great variations in the symmetry exist, not only in different genera of the same order, but in different species of the same genus.*

Now, with reference to this latter fact, it must be borne in mind that flowers are so highly differentiated from the

---

* See note by the author, *On the Causes of the Numerical Increase of Parts of Plants*, Journ. Lin. Soc. Bot., xvi. p. 1.

THE PRINCIPLE OF NUMBER. 13

leaf type, that they have undergone such wonderful transformations and adaptations to insect and other agencies and to their environing conditions, so that the simple and original laws governing the arrangement of the leaves, here propounded for the origin of what may be called the "primitive symmetry" of the floral organs, have become in many cases masked or interfered with. Hence, to deduce those original laws from the present structure of flowers, it is not only necessary to consider the floral symmetry of an immense number of genera, and so ascertain what are the relative proportions of certain numbers when associated with alternate and opposite leaves respectively, but to discover what may have been the interfering causes which have modified what would have been the immediate effects of the fundamental laws of phyllotaxis.

Thus, it will be found that the numbers of the parts of whorls are liable to vary on their own account, while the arrangement of the foliage varies independently at the same time; so that where the floral symmetry of a plant does not tally with the leaf arrangement, the discrepancy may be due either to subsequent changes occurring in the flowers or in the leaves, or perhaps in both.

For example, a quaternary floral type may be, and often is, associated with alternate leaves; where there is reason to suspect that the former was established from a primitive opposition in the leaf organs, but that the foliage has subsequently differentiated into a spiral arrangement, leaving the original 4-merous symmetry of the flowers unaffected, as in many of the *Onagraceæ*; *Epilobium*, indeed, often furnishing ocular demonstration, as, while the lower leaves may be opposite, the upper are often alternate.

On the other hand a quinary arrangement is often associated with what may be called a persistent opposition in the leaves, as in *Caryophylleæ* and *Labiatæ*. This may be due either to an abrupt change from opposite leaves or bracts to a spiral one in the flower, or by a reversion from an alternate to an opposite position of the leaves, the floral organs retaining the arrangements due to their spiral origin.

The symmetry is based on Calyx, Corolla, and in many cases the Andrœcium also; but the carpels are not generally regarded, for it does not usually extend to the gynœcium, though it is very frequently retained in the andrœcium, which is often some multiple of that of the perianth whorls.

In presenting the reader with what may be regarded as ostensible grounds for the interpretation proposed, attention will be first directed to the more obvious correlations between floral symmetry and leaf arrangements, as appear from certain numerical proportions; and, in the next chapter, to significant facts observable in the symmetry of particular plants.

- Commencing with genera possessing alternate leaves and a quinary floral type, the prominent fact becomes at once apparent that this correlation far exceeds in numerical proportion any other. Thus, of above eighty Dicotyledonous orders [*] examined in all, no less than 1285 genera have quinary flowers associated with alternate leaves, and this is exactly what one would expect according to the theory advanced that 5-merous whorls are cycles of the $\frac{2}{5}$ type.

As a corroboration is the fact that such whorls often have their parts arranged quincuncially in æstivation (Fig. 3, *a*); and when they are not so they can be referred to

[*] I consulted the first volume of the *Genera Plantarum* for this purpose, which embraces the *Thalamifloræ* and *Calycifloræ*.

THE PRINCIPLE OF NUMBER. 15

it, as I have explained elsewhere : * thus Fig. 3 shows how the varieties of imbricate æstivations are deducible from the ⅖ type (*a*), by shifting the edge of the 2nd member under the 4th (*b*, "vexillary"), the 3rd under the 5th (*c*, "imbricate proper"), and the 1st under the 3rd (*d*, "contorted").

Similarly ternary or trimerous whorls are almost universal amongst flowers of Monocotyledons, and the ⅓ type of phyllotaxis is equally common in the foliage. It has been seen that the ⅓ type cannot be deduced from opposite leaves, and consequently never occurs, as far as I know, amongst the foliage of Dicotyledons. The comparatively few genera in this class with ternary flowers is therefore in accordance with the views herein expressed; and where they occur, as

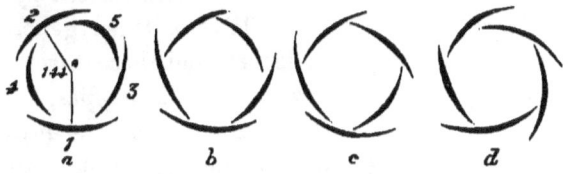

Fig. 3.—Floral Æstivations.

in *Berberis*, there are special features which lead one to believe they are not due to the ⅓ type at all, but to the breaking up of a high continuous spiral into groups of threes, as will be explained hereafter.

If, however, we take a theoretical departure from a single cotyledon, as occurs in Monocotyledons, then the next leaf can be at either of the limiting positions of the angular distances of 180° or 120°, but not less; for if it were less than 120°, there would be four leaves in any projected circle, and this would immediately introduce a member of the series ⅕, ¼, ²⁄₇, etc., as shown above. The consequence is

* See my paper, *On the Origin of Floral Æstivations*, Trans. Lin. Soc., 2nd series, BOTANY, vol. i. p. 177.

that the $\frac{1}{2}$ and $\frac{1}{3}$ types are exceedingly common in the foliage of Monocotyledons, while the $\frac{2}{5}$, as far as I am aware, is entirely wanting in that class, whether in foliage or flowers.

Of genera having alternate leaves but associated with a binary or quaternary floral symmetry, there are about 270 in number of about 30 orders. Now, the co-existence of alternate leaves with 2- or 4-merous flowers appears at first sight to negative the theory; but, as mentioned above, these and other irregularities have been brought about by subsequent differentiations in the foliage or flowers. On the other hand, opposite leaves with quaternary flowers are not at all infrequent, though not quite so common as when they are alternate; thus, *Oleaceæ* and *Onagraceæ* are so conditioned. Again, in *Rosaceæ*, which is an order characterized by having alternate leaves and 5-merous flowers, three genera alone out of seventy have opposite leaves, and these three also are accompanied by 4-merous flowers; viz. *Rhodotypus*, *Coleogyne*, and *Eucryphia*. These three genera thus acquire their importance from being isolated amongst others to which they are allied, and which are generally otherwise characterized. Many orders have both foliage and floral symmetry remarkably inconstant, and all four combinations, viz. 4-merous and 5-merous flowers with opposite or alternate leaves almost indiscriminately, as in the tribes *Diosmeæ* and *Borosmeæ* of *Rutaceæ*; and it is a noticeable fact that, associated with this inconstancy of correlation, there is an inconstancy in the leaf arrangement, opposite and alternate leaves being often in species of the same genus, and even on the same individual plant.

The total number of genera noticed as having 4-merous flowers and opposite leaves was 110 in 25 orders; whereas I noticed 276 genera of 30 orders as having 4-merous flowers associated with alternate leaves. This, I believe, is due to subsequent differentiation in the foliage to an alternate condition, the quaternary condition of the flowers remaining unaltered.

Similarly with the last condition, I found 212 genera of 30 orders with a quinary arrangement of the flowers correlated to an opposite condition of the leaves, this being an apparent anomaly of the same kind, but which is, however, to be interpreted in the same way. Thus the *Labiatæ* are constantly 5-merous in the flowers, but with as constantly opposite leaves. Now, if we contrast this order with *Scrophularineæ*, we find a similar constancy in certain genera only, as in *Rhinanthus*, etc.; while other genera have alternate leaves as *Linaria*, *Digitalis*, etc.

There is an alternative of interpretations of this fact, for both can be illustrated in nature. Either all the pentamerous flowers have been deduced from alternate leaves (as may have been the case with *Rhinanthus* and *Labiatæ*), the leaves having subsequently reverted to the original or ancestral state of opposition; or else, the 5-merous character of the flowers has arisen by *a sudden change* (possibly due to the stimulus of insect agency) from opposition in the leaves or bracts to an alternate arrangement in the parts of the flower. As an illustration of this latter process may be mentioned the development of the five sepals of *Deutzia* as compared with the four of the allied genus *Philadelphus*. In this latter genus the anterior and posterior sepals appear together, subsequently the two lateral arise simultaneously. In *Deutzia*, however, the two anterior sepals correspond to Nos. 1 and 3; two sepals are lateral, viz., Nos. 4 and 5; and the posterior sepal is No. 2. Thus the opposite and decussate pairs of sepals of *Philadelphus* would be represented by the figures 1 and 2, 3 and 4. If these were to break up into a quincuncial spiral and shift their positions.

they would, with the addition of one more sepal, assume those represented by *Deutzia*.

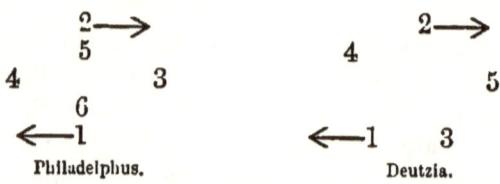

Exactly the same procedure occurs in the change from opposite to alternate arrangements of leaves in the Jerusalem Artichoke, as I have explained in treating of the varieties of leaf-arrangement in that plant.

Calycanthus is another instance illustrating an abrupt change from an opposite condition of the leaves to the $\tfrac{2}{5}$ type in the bracts enveloping the flowers, and which then pass insensibly into sepals and petals.

SYMMETRICAL INCREASE AND DECREASE IN FLORAL WHORLS.—As another instance of variability adding further complications, it may be observed that in both kinds of arrangements, namely, of those plants possessing alternate and those possessing opposite leaves, there are many genera whose floral symmetry ranges from one to some higher number in the different species of the same genus. Thus 4-5-merous flowers are especially common. I found it so in more than 100 genera of 23 orders examined among alternate-leaved plants; and 58 genera of 19 orders among those with opposite leaves.

Again, some genera have species the whorls of whose flowers range from 3 to 5 or 6, or from 4 to 6 in the number of parts; others from 5 to 7 or 5 to 8, etc. In these cases it is often quite impossible to explain what has been the immediate causes producing such variations. The only interpretation that can be given is that the primary symmetry having been originally determined by phyllotaxis, it changes, whether in the individual or in its descendants, through the law of "symmetrical increase or decrease." By this I mean that the number of sepals, petals, and stamens often vary together from the typical number by the addition or subtraction of a member. Thus, in a single corymb of an Elder, 4-, 5-, 6-merous flowers may be often found; similarly, while early blossoming Fuchsias may bear 3-merous flowers, they are replaced later by the regularly 4-merous ones. Although these changes frequently occur in the same plant, they usually are not permanent. Yet they occasionally appear to have become so, as in the terminal flowers of *Adoxa* and *Monotropa*. On the other hand, the constant occurrence and, therefore, specific character of 4-merous flowers in *Potentilla Tormentilla*, and 3-merous in *Tillæa muscosa*, I should be inclined to attribute to the fixation of a symmetrical reduction which has taken place from the permanent 5-merous type so characteristic of *Potentilla*, and many genera of the *Crassulaceæ*. Not infrequently the difference of number is pronounced by systematists as generic; thus, while *Rubia* has 5-merous flowers, *Galium* has 4-merous. A similar difference lies between *Ruta* and *Haplophyllum*.[*]

If a cause be looked for, it would seem to be merely a question of nutrition. If the symmetry varies in the same plant, it is obvious that a corolla of four petals could not have been provided with the same amount of nutritive material as a 5-merous one. But if it be a specific character, as in Tormentil (which, it may be observed, affects the more or less barren soil of heaths), then the change has become fixed and is now hereditary.

[*] By running the eye through the artificial keys at the commencement of the Orders in the *Genera Plantarum* of Bentham and Hooker, it will be seen how frequently these authors regard the number of parts in the Calyx and Corolla as a prominent generic character.

UNSYMMETRICAL DECREASE IN CERTAIN FLORAL WHORLS.—
Another modifying cause of the change of symmetry is
the adaptation to insect or other agency for fertilisation.
This I believe to have played a most important part in modifying flowers, as will be explained more fully hereafter, more
especially in affecting the Andrœcium and Gynœcium, than
the Perianth, as far as "number" is concerned, this latter
organ being altered by their agency, more especially in Form.
Thus, the loss of one or more stamens is very characteristic of
certain groups, as in the *Labiatæ*, when the remaining members of the andrœcium become altered in length and position
so as to facilitate the intercrossing of distinct flowers.

On the other hand, with inconspicuous and cleistogamous
flowers, there is a strong tendency to reduce the number of
stamens, as in Chickweed to three, the allied species *Stellaria
Holostea* having ten. Similarly, in the cleistogamous flowers
of Violets they are sometimes reduced to three or two; since
a very small amount of pollen is really quite sufficient to
fertilise a considerable number of ovules.

The gynœcium has very frequently a less number of
carpels than the other whorls have parts. Now, the primary
effect of intercrossing is to enhance the size of the corolla
and to give a preponderance to the andrœcium. On the
other hand, one result is to check for a time the growth and
development of the gynœcium of most insect-visited hermaphrodite flowers, *i.e.* to render the flower protandrous; and
I strongly suspect that the generally reduced number of
carpels in highly differentiated flowers—as of the *Gamopetalæ*,
in comparison with the *Thalamiflorœ* and *Calyciflorœ*—is correlated to the fact that they have been for many generations
visited by insects. This idea is supported by the fact that
bicarpellary genera sometimes tend to restore the ancestral
number of the five carpels, as is occasionally the case in
Gesneria.

THE PRINCIPLE OF NUMBER.

In some cases, nature seems, as it were, to try and compensate for the loss of the carpels by an increase in the quantity of seeds. Thus, while no Labiate flower has more than four seeds, it has been ascertained that a *Maxillaria* bore 1,700,000 seeds; and I found by calculation that a single plant of Foxglove yielded a million and a half apparently good seeds.

The relative advantages of having many or few seeds will be discussed later on.

ILLUSTRATIONS FROM RANUNCULACEÆ.—Certain genera of the *Ranunculaceæ* are particularly instructive in showing how members of the floral whorls originate in phyllotactical methods, but are more or less altered in their positions by the lateral union of their fibro-vascular cords; so that they become arranged in superposition instead of being alternate, or *vice versâ*. Thus, in *Garidella* (Fig. 4) (with which *Helleborus fœtidus* partly agrees), the sepals and petals are both arranged, and arise successively, in quincuncial order; the petals being (correctly, in accordance with phyllotaxis) superposed to the sepals.

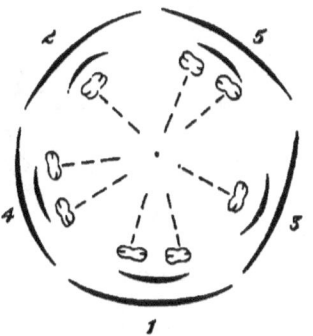

Fig. 4.—Diagram of Garidella.

The andrœcium forms a whorl of eight stamens, and represents a cycle of the ⅜ arrangement; the proper angular divergence of 135° is, however, not retained, in consequence of the fibro-vascular cords being intimately connected with those of the petals. Having thus established the first whorl of eight, the rest of the staminal series follow on the same radial lines. By referring to the diagram (Fig. 4) it will be seen how the stamens of the outermost whorl group themselves in superposition to the petals and sepals. Similarly, in *Nigella sativa* the petals are eight in number, and occupy the same positions as the outermost whorl of stamens of *Garidella*. They have, then, the eight stamens of the outermost whorl of the andrœcium superposed to them.

In *Delphinium* the stamens and carpels form a continuous spiral, represented by $\frac{3}{8}$, or approximately by $\frac{3}{8}$. In some cases Braun* found 16 stamens, and the first carpel being the 17th organ, stood superposed to the stamens No. 9 and No. 1. In another case 18 stamens were developed, so that the first carpel stood superposed to stamen No. 11.

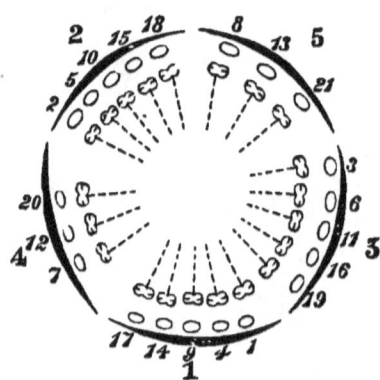

Fig. 5.— Diagram of *Helleborus niger*.

Helleborus niger (Fig. 5) has five sepals which emerge and are arranged in quincuncial order. There are twenty-one nectariform petals, *i.e.* one cycle of the $\frac{8}{21}$ arrangement, grouped as in the accompanying diagram. The petals 1 to 8 and 9 to 16 would correspond approximately to two cycles of the $\frac{3}{8}$ type. Radial rows of stamens then follow on the same lines as the petals.

Eranthis hyemalis has, as usually regarded, a 5-8-merous coloured calyx. A pair of staminodes stand superposed to each member of the outer whorl. Stamens follow along the radial lines, of which six terminate in carpels.

Aquilegia vulgaris, or the Columbine, has the sepals, as

* Al. Braun on *Delphinium* (Pringsheim's *Jahrb. f. Wiss. Bot.*, 1857, i. 206), referred to by Henfrey, *Morphol. of Balsamineæ*, Journ. of Lin. Soc., iii. 159.

usual, quincuncially arranged. The petals appear *simultaneously*, alternating in position to the sepals. The stamens occur in ten rows, 5 being superposed to the petals and 5 to the sepals; and, lastly, 5 carpels appear superposed to the petals. This flower, then, adopts the more usual character of alternation in the whorls. But it may be noticed that while the corolla alternates with the calyx, each of these outer whorls gives rise to a radial series of stamens.

From the preceding illustrations, it will now be seen that phyllotaxis lies at the foundation of the arrangements of the members of floral whorls; that the $\frac{2}{5}$ type prevails in the sepals and petals, with a strict angular divergence of 144°. The divergences are, however, subsequently modified in the stamens and carpels. Thus, in *Helleborus niger* the petals clearly represent a whorl of 21 parts, *i.e.* they are presumably arranged according to the $\frac{8}{21}$ type. They are, however, so far modified in position as to become superposed to the sepals in groups. Similarly the stamens form series of 21, each being superposed in radial lines to the petals.

The interpretation of these displacements from what would be due to strict, phyllotactical laws is that the individual cords of the stamens and carpels are not independent as they are in the "leaf traces" of an axial cylinder, where the cord or cords belonging to each leaf are simply intercalated side by side with those of the leaves most nearly approaching the same vertical line, and constitute together the common fibro-vascular cylinder of the stem. In the pedicel, however, the rule is that this should contain at least the same number of cords as there are leaves to the perianth, or sepals and petals together. These, usually six or ten cords, on reaching the floral receptacle are sent off respectively as the cords of the sepals and petals; whereas, it is *these latter* which by lateral or radial "chorisis" supply the cords required for the stamens and carpels

The consequence is that the essential organs have their cords issuing from a common stem with those of the perianth. Thus they are *compelled* to stand superposed to them.

Perhaps the word "compelled" requires a word of explanation. The cord of any organ superposed to another may be given off either by radial, *i.e.* lateral, or tangential chorisis from the cord of the latter. Instead, however, of the new lateral branch giving rise to an organ by the side of the former, it results, partly from the close proximity of the two and partly from the tendency of the remaining cords of the cylinder to "close up," that the new member finally takes up a position in front of, *i.e.* superposed to, the one whose cord has given rise to it. When a cord is separated by tangential chorisis, as is so often the case with staminal cords, then the resulting organ must necessarily be superposed to the one, from the cord of which it has been detached.

CHAPTER III.

THE PRINCIPLE OF NUMBER—*Continued*.

ILLUSTRATIONS OF SPECIAL NUMBERS.—It will now be advisable to give examples of particular numbers occurring in flowers, and attempt to account for them.

ONE-MEMBERED WHORLS.—Where one part to a whorl is only found, it may in nearly every case be regarded as a degradation from some higher number. The only instances I am aware of in which the calyx seems to consist of a single member are some species of *Aristolochia*. In *Mussænda* one out of the five sepals is greatly enlarged to become an attractive organ.*

One petal is occasionally found. Thus, four genera of *Vochysiaceæ* have each only one petal to their flowers; but as the sepals are five in each of the seven genera of this order, and the petals range from one to five in number, the inference is clear that the solitary petal of these four genera is due to the arrest of the others.

One stamen occurs more frequently; as in *Hippuris, Centranthus, Euphorbia, Casuarina, Orchis, Canna, Lilæa, Lemna*, etc. As allied genera have more than one, and it is accompanied by other signs of degradation or metamorphosis,

* If there be one external foliar organ only, it is regarded as a bract, as in Willows and *Aponogeton*.

there is no doubt but that similar processes will account for one stamen as for one petal. Thus *Hippuris* with one, is allied to *Myriophyllum* with four; while *Centranthus* has one, *Fedia* has two and *Valeriana* three.

Casuarina alone seems to raise a doubt of its being degraded and possibly a primitive form; but this is solely because it has no living allies (excepting perhaps *Myrica*). The terminal stamen would not be of itself a point of importance, as it has a parallel in *Euphorbia;* but it is its isolation without affinities, its peculiar equisetum-like habit, which seem to indicate great antiquity, so that no inference can fairly be drawn to interpret its present monandrous condition.

Amongst Monocotyledons, *Canna* is clearly monandrous by petalody of the other stamens, *Orchis* by metamorphosis also. Lastly, *Naias, Caulinia, Zostera, Zannichiella,* and *Lemna* are in all probability greatly degraded forms from higher plants, degradations being the usual effect of an aquatic life, and not primitive types of Monocotyledons.

One carpel is not at all uncommon, as in the *Leguminosæ*. As *Affonsea* has five, the absence of four in this order is no doubt due to arrest. In the tribe *Bereræ*, however (if my interpretation be correct, of the origin of the seven whorls of three each constituting the flowers of *Berberis*, as explained below), the one carpel may be the last of an originally continuous spiral, formed from eleven pairs of opposite leaves, now broken up into seven ternary whorls, with one over. It may, however, be the remaining one of three, which possibly constitutes a ternary gynœcial whorl, which is characteristic of the tribe *Lardizabaleæ* of the same order *Berberideæ*.

DIMEROUS WHORLS.—A dimerous arrangement is not particularly common, though a quaternary calyx is dimerous in its development, as the sepals emerge from the axis in successive pairs.* The following may be taken as illustrative instances. The sepals of *Papaver* and *Fumaria,* the outer stamens of *Cruciferæ.* In *Circæa* all the whorls are dimerous, in *Oleaceæ* the essential organs alone, as also in *Pinguicula, Salvia, Veronica,* and *Salix diandra.*

The question arises, is this number two an original one, or has it arisen by arresting some parts of a more numerous whorl? It is obviously so with *Salvia* and other genera of the *Labiatæ,* where rudimentary stamens are present. So also with *Senebiera didyma* where the two stamens take the place of the four larger ones of other genera of the *Cruciferæ.* It is probably so with the two imbricate sepals of Poppies, those of *P. orientale* being often increased to three, which seems to be a tendency to revert to a more primitive and higher number.

With such plants, however, as *Circæa,* the Ash, and *Veronica,* which have retained opposite leaves, the dimerous whorls *may* be a primitive condition. This idea is ostensibly supported by the fact that the outer whorls of the flowers are quaternary and not quinary, since, when this is the case, the sepals always issue in pairs from the axis, and not simultaneously as do the petals; but as long as no rudimentary organs exist, there is nothing to disprove the idea that in these genera the number of stamens may not be due to degradation. Indeed, all analogy would lead one to suppose so in most cases, as of *Circæa* and *Veronica:* the binary whorls of the former genus, and the quaternary outer and binary inner whorls of the latter, being presumably due to "symmetrical reduction" from the prevailing quaternary

* Though the antero-posterior sepals of cruciferous flowers are regarded as the most external, it is really the lateral ones which are first provided with fibro-vascular cords from the complete oblong cylinder in the pedicel, just as in *Cleome* (see Fig. 6, p. 32).

type of the *Onagraceæ* and quinary of the *Scrophularineæ* respectively.

TRIMEROUS WHORLS.—The number three is strongly characteristic of Monocotyledons, and appears to be in this class the immediate result of the $\frac{1}{3}$ phyllotaxis. In Dicotyledons, however, there are certain orders in which it prevails, and it will be noticed that the number of parts in those orders is generally much increased; as in *Magnoliaceæ*,* *Anonaceæ*, *Berberis*, *Laurus Camphora*, *Rumex*, etc. In some the andrœcium and gynœcium are so increased in number that they cease to be whorled, but have become spirally arranged on a more or less elongated receptacle and are represented by the fractions $\frac{5}{13}$ or $\frac{8}{21}$.*

It has been demonstrated above that a pentamerous arrangement is undoubtedly due to the $\frac{2}{5}$ phyllotaxis, each whorl constituting a cycle; but if the fraction be a higher one, as $\frac{5}{13}$ or $\frac{8}{21}$, then the number of parts in a cycle are too great to be compressed into a whorl. Nature appears then to adopt another method. Falling back upon the law that with these arrangements no part of the continuous spiral, of sufficient length to constitute a complete circle when projected upon a plane, ever contains more or less than three leaves (excepting the $\frac{1}{3}$ type), the series is now broken up into a succession of ternary whorls, the whole forming the complete flower, and, being taken together, corresponds to about or exactly one cycle of a high type. Thus Barberry has 3 bracts, 3 + 3 sepals, 3 + 3 petals, 3 + 3 stamens and one carpel; that is, *seven whorls* of threes or *twenty-one*

---

* In *Magnolia* an individual complication is introduced, in that the immense number of stamens and carpels is secured by doubling the whole number attributable to the $\frac{5}{13}$ arrangement. Consequently, instead of there being *five* and *eight* "secondary spirals," there are *ten* in one direction and *sixteen* in the other.

parts, and one over. If these seven whorls were broken up and arranged spirally, they would be represented by $\frac{8}{21}$; and then there would be *eight coils* in the cycle. The presence of seven and not eight whorls is due to the fact that in rearranging them, so to say, in a verticillate manner, and by necessarily shifting the position of the parts, a certain portion of the spiral line is lost in forming each whorl, as the angular divergence between two parts in a whorl is 120°, but on the spiral it is nearly 123°; so that by the time the twenty-first organ is arrived at, only seven circles have been completed.

Similarly, in *Rumex*, if we supply the theoretically lost corolla, the flower would consist of twenty-one parts exactly.*

Another and somewhat frequent origin of the number three in Dicotyledons is due to what I have called symmetrical reduction: when not only the different species of a genus may have the number of parts of their floral whorls ranging from 5 to 4 or 3; but such variations may occur on the same plant. Thus *Rutaceæ* (following the *Gen. Plant.*) has 34 genera with 5-merous flowers; 18 genera with species varying from 5 to 4-merous; 16 are 4-merous; 3 range from 5 to 3-merous; 2 from 4 to 3-merous, and 1 is 3-merous.

TETRAMEROUS WHORLS.—That a true quaternary arrangement is due to an opposite condition of the foliage seems borne out by statistics, though quinary flowers are not at all uncommon as well. Thus of *Rutaceæ* there are 6 genera with opposite leaves and 4-merous flowers; 2 only with 5-merous, and 2 with 4-5-merous flowers. On the other hand, there are 25 genera with alternate leaves and 5-merous flowers.

* High spirals can be otherwise treated, as in the case of *Chimonanthus*, where whorls of fives are made out of a spiral system of $\frac{8}{21}$ (see below, p. 38).

Another correlation with a quaternary arrangement is a not unfrequent valvate condition of the sepals at least, or of the sepals and petals as well. These conditions prevail, for example, in *Oleaceæ, Onagraceæ*, and, with the exceptional genus, *Clematis*, of the *Ranunculaceæ*. Too much stress must not be placed upon this coincidence, as, if the petals be enlarged through insect or other agency, the valvate æstivation is often lost, and the petals become imbricate, as in *Fuchsia, Godetia*, etc., though it is there retained in the sepals. This valvate condition is foreshadowed in the vernation of the foliage; in that opposite leaves are almost invariably valvate, having the two upper surfaces of the leaves pressed together, as may be seen in *Hypericum* and *Vinca*; or else with the edges induplicate, as is characteristic of *Caprifoliaceæ*, resembling the sepals of *Clematis*.*

Though the *Onagraceæ* have a preponderance of genera with 4-merous flowers, there is in this order great variation in the foliage. It is strictly opposite in *Fuchsia* and others, but 14 genera out of a total of 22 have alternate leaves, while with some, like *Epilobium*, it varies on the same stem. This, I think, reveals the fact that the 4-merous condition has been first established in the flowers, and subsequently the foliage has varied from an opposite to an alternate condition in certain genera, just as it does in an individual plant of *Epilobium*.

That symmetrical reduction has elsewhere played an important part in the origin of 4-merous flowers, is a supposition fully borne out by facts. In some cases it has seemingly established itself as a permanent character, so that systematists recognize it as generic or specific, accordingly,

---

* See a paper by the author, *On Vernation and the Methods of Development of Foliage as protective against Radiation*, Journ. Lin. Soc. Bot., vol. xxi., p. 624.

THE PRINCIPLE OF NUMBER. 31

as the case may be; this, *Haplophyllum* may be compared with *Ruta*, *Rubia* with *Galium*, or, again *Potentilla reptans* with *P. Tormentilla*, etc. On the other hand, I repeat, when one observes that of the 71 genera of *Rosaceæ* three only are recorded in the *Gen. Plant.* as having opposite leaves, and these three are characterized as having 4-merous flowers, viz. *Rhodotypus*, *Eucryphia*, and *Coleogyne*, there appears to be a significant correlation between quaternary flowers and opposite leaves.

A quaternary arrangement is found very exceptionally in Monocotyledons, as in the order *Naiadaceæ*, e.g. *Tetroncium* and *Potamogeton*. As the numbers 6 (*i.e.* 2 × 3), 4, 2, and 1 are found in different genera, the quaternary as also binary arrangements may, I think, be reasonably referred to symmetrical reduction.

Perhaps of all orders the quaternary arrangement (at least in part) of Crucifers has raised more discussion than any other kind of floral symmetry.*

Without entering here upon any lengthened discussion I would only add that, as far as investigations into the anatomical structure of the pedicel is concerned, there is a decided difference from what occurs in most flowers having a definite number of parts, and where the whorls are regularly superposed to one another, in that the members of the whorls not being for the most part on common radial planes, they have not their cords fused together in the usual manner in a *radial* direction.

A section at some distance below the flower reveals four or five cords forming a circle. These rapidly increase in number by branching laterally, till between ten and twenty are found arranged in an oval just below the flower. Two

---

* See my paper *On the Structure of a Cruciferous Flower*, Trans. Lin. Soc., 2nd series, BOTANY, vol. i. p. 191.

cords, one at each end of the long axis, now part company from the rest, and enter the lateral sepals (Fig. 6 (a) *l.s.*), the antero-posterior sepals next receiving their cords (*a.s.* and *p.s.*). The cylinder tends to close up, and four groups situate at the corners of the oblong cylinder supply cords for the petals, *p.* The two honey-glands next put in an appearance, *G.* They are merely cellular expansions of the floral receptacle, and are entirely devoid of cords, and therefore not rudiments of appendages. The two lateral stamens next receive their cords, *l.st.*, while four other cords are given off from beside the petaline for the taller pairs of stamens, *st.*

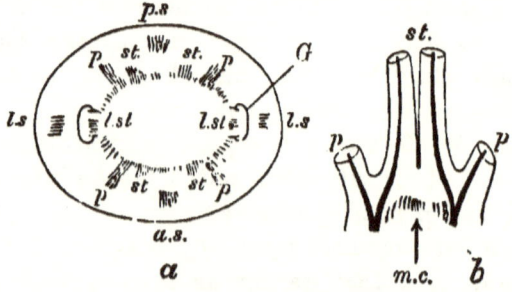

Fig. 6.—Anatomy of Wallflower.

Fig. 6 (*b*) shows how their cords *diverge* below and spring from the side of the petaline cords, while extra cords arise between them to form the marginal cords of the carpels (*m.c.*). From this it will be seen that the longer stamens cannot be formed by "chorisis" of a common intermediate cord; but, like those of all other members of the flower, their cords are separated from the common fibro-vascular cylinder of the stem.

The conclusion suggested by this investigation, and by a comparative study of *Capparideæ*, is that a cruciferous flower is not reducible to an originally quaternary type at all, but to some higher one. In my paper referred to, I suggested a quinary; but I am now more inclined to refer it primarily to an indefinite spiral series referable to the $\frac{5}{13}$ or $\frac{8}{21}$ type, which has been reduced, perhaps through insect agency, by symmetrical reduction to the present anomalous condition.

The process of transition from a hypothetical indefinite number of stamens to the present hexandrous state may be, perhaps, seen by comparing the three genera of *Capparideæ*—*Capparis, Polanisia and Cleome.* The first has many stamens and six placentas, which are sometimes reduced to two. *Polanisia* has eight stamens, or more rarely six. Their situations correspond exactly with those of the *Cruciferæ*, except that, when there are eight, there are four on the anterior side instead of two.

Lastly, *Cleome* brings us to the same structure as in the *Cruciferæ* with even the tetradynamous condition of the stamens; the elongated torus below the pistil being about the only "capparidaceous" feature left.

It is not at all uncommon to find more than six stamens in cultivated plants of the *Cruciferæ*, and when this is the case I should be inclined to regard it as a tendency to a reversion to a higher ancestral number.

On the other hand, the close proximity of the two taller ones on each side not infrequently brings about some degree of cohesion between them, with an occasional arrest of half an anther. This has led some to suppose that the pair have resulted from chorisis. Since, however, their cords *diverge* downwards to the right and left, and run down beside the petalline cords(Fig. 6, *b*), this clearly proves that the union is a result of close contact, and that the normal separation is not due to chorisis, but to a primitive freedom, which has been retained from a multistaminate condition.

PENTAMEROUS WHORLS.—These are by far the commonest amongst Dicotyledons. And as an enormously greater proportion of plants in this class have alternate leaves and 5-merous flowers, this correlation alone would be almost sufficient to prove that the latter issued out of the commonest or $\frac{2}{5}$ type of phyllotaxis. But since the sepals are sometimes decidedly quincuncial, as are those of *Digitalis*, and the petals frequently so, we have undoubted proof that they represent cycles of this angular divergence.

As with other numbers, fives may arise by symmetrical increase from fours, or decrease from sixes; though in by far the greater number of instances it is a primitive number, as stated above. As a rare instance of symmetrical decrease may be mentioned *Lythrum Salicaria*, which has usually the central floret of each axillary cyme 6-merous, but the lateral ones only 5-merous. As an instance of five parts to a whorl amongst Monocotyledons, may be mentioned the stamens of *Strelitzia regina;* but this number is obviously due to the suppression of a stamen.

Although whorls of fives are cycles of the $\frac{2}{5}$ divergence, and usually follow after an alternate arrangement in the foliage, yet it is quite possible to change abruptly from opposite leaves or bracts to whorls of fives in the flower, as may be seen in *Hypericum* and *Dianthus*. This arrangement, as I have elsewhere shown, is that most easily acquired when opposite and decussate leaves become alternate by the development of internodes (see pp. 11 and 18).

HEXAMEROUS WHORLS.—A floral whorl of six parts is, in most cases, as amongst Monocotyledons, the result of the combination of two whorls of three each—as the andrœcium of *Berberis*, Tulip, or perianth of the Lily of the Valley. It may, however, arise from symmetrical increase, as, for example, in the orders *Meliaceæ* and *Olacineæ*. In the former, there are 18 genera with alternate leaves and 5-merous flowers; 9 with 4-5-merous; 4 with 4-merous;

4 with 5-6-merous, and 1 with 4-6-merous whorls in the different species. In *Olacineæ*, of 36 genera, 17 have alternate leaves and 5-merous flowers; 7 have 4-5-merous; 4, 5-6-merous; 2, 6-merous, and 1, 4-6-merous.

As six leaves cannot form a cycle of any of the ordinary kinds of phyllotaxis, this will account for its rarity in nature; and indeed it may probably, without exception, be divisible into two whorls of three members each, except in the case of symmetrical increase from five.

HEPTAMEROUS WHORLS.—Like the number 6, 7 is a very rare one; and when present appears to be due to its being a primitive number or to symmetrical change. If any whorls are deducible from decussating verticils of threes, a cycle may contain seven parts, as the phyllotactical series arising from the breaking up of such verticils into a continuous spiral arrangement is represented by $\frac{1}{3}$, $\frac{1}{4}$, $\frac{2}{7}$, $\frac{3}{11}$, etc. So that if leaves on a plant were in whorls of threes, as occurs in some instances, and not opposite, as in the primitive type amongst Dicotyledons, then a heptamerous arrangement would occur. If, therefore, there be any existing illustration, it must, by the very nature of the case, be exceedingly rare. It sometimes occurs in *Trientalis;* and when this is the case, it may possibly have arisen as here suggested. According to the description given of this plant in the *Genera Plantarum*, the numbers of the three outer whorls range from 5 to 9, the capsule being 5-valved. The leaves, on the other hand, are "sæpe tot quot petala subverticillata."

A second cause is arrest. This obviously accounts for the 7 anthers in *Pelargonium*, for the 10 filaments are present.

A third cause is symmetrical change. *Lythrum Salicaria* illustrates this as already mentioned. This flower is sometimes described as 6-merous, but it is not always so. The central floret of the cyme has often a higher number than that of the lateral ones; so that if they be 6-merous, the central flower will be 7-merous. *Agapanthus*, amongst Monocotyledons, is another instance, its flowers ranging from 6 to 8 in the number of parts in the whorls.

OCTAMEROUS WHORLS.—A whorl of eight parts is not common; but it appears in *Chlora* and in the corolla of *Dryas octopetala*, in which it *may* be a cycle of the $\frac{3}{8}$ phyllotaxis. In other cases it is a combination of two whorls, which, as a rule, can be easily distinguished as the stamens in the *Onagraceæ*, or it may be due to symmetrical change.

ENNEAMEROUS WHORLS.—The number 9, like 6, 7, and 11, corresponds to no cycle of any one of the usual forms of leaf-arrangement, and is proportionately rare. It may occur as a combination of three cycles of three each, and perhaps this will account for it when it occurs in *Trientalis*, and the androecium of *Mercurialis*. The stamens of *Butomus* are also nine in number.

DECAMEROUS WHORLS.—The number 10 never occurs except as the union of two whorls of five in each, as in the androecium of *Leguminosæ*.

ENDECAMEROUS WHORLS.—Like 7, the number 11 might occur if the series $\frac{1}{3}$, $\frac{1}{4}$, $\frac{2}{7}$, $\frac{3}{11}$, etc., was as frequently represented as $\frac{1}{2}$, $\frac{1}{3}$, $\frac{2}{5}$, $\frac{3}{8}$, etc., when "sevens" would be as abundant as "fives" are now. I do not know of a case where it could reasonably be referred to such an origin. When it does occur, as in *Cuphea*, it is clearly due to an arrest of one stamen through insect agency. *Brownea* is said also to have sometimes 11 stamens; if so, this would undoubtedly be due to numerical increase.

DODECAMEROUS WHORLS.—The number 12 closely verges on the "indefinite," which simply means a more or less numerous series of cycles of the same kind. Neverthe-

THE PRINCIPLE OF NUMBER. 37

less, it occurs as a "definite" number in several instances. The 12 stamens of *Lythrum* are, of course, two series of six each. Both 12 and 24 are found in the *Crassulaceæ*, as in *Sempervivum*, in which genus the petals vary from 6 to 20, and the stamens from 12 to 40. This seems to show that in the one case they are combinations of cycles of threes, in the other, of fives; just as *Berberis* illustrates the former, *Chimonanthus* the latter instance.

INDEFINITE WHORLS.—As soon as we pass from twelve to some higher number, then flowers cease to be whorled, and the parts are arranged spirally, and follow more or less exactly the laws of alternate phyllotaxis; interferences occur in consequence of the want of space, some secondary spirals being often incomplete. Moreover, since the fibrovascular cords become fused, in other words branch by chorisis, and are not independent as of ordinary foliage, parts take up slightly different positions to what they would if they could strictly follow phyllotactical laws.

I have alluded to what I call "symmetrical increase and decrease" as causes of variation in the number of parts of whorls; and what brings about these variations in number, is an excess or deficiency of nutriment and vital activity respectively. There are innumerable examples of all the above kinds of changes in number. In fact, if any one or series of whorls of a flower be n-merous, it may become $n \pm x$-merous, and will give rise to symmetrical increase or decrease accordingly; or again, three whorls of the same flower may become $n \pm x$, $n \pm y$, $n \pm z$-merous; when all numerical symmetry between them will be destroyed.

Similarly, if the parts be spirally arranged, the number may vary from the prevailing one by increasing or decreasing the length of the spiral, both in flowers of the same plant or in different species of the same genus; as, for example, may be seen by comparing the number of stamens in a large-flowered form of *Ranunculus aquatilis*, with the small-flowered *R. hederaceus*; or one genus with an allied one, as *Ranunculus* with *Myosurus*, in which the stamens are reduced, often to one whorl of five only.

Lastly, just as high spirals can be broken up into ternary whorls, so can the arrangement $\frac{8}{21}$ be separated into whorls of a lower series, as of 13, 8, or 5 parts respectively. Thus, of the two genera, which have opposite leaves, comprising the order *Calycanthaceæ*, *Calycanthus* illustrates an abrupt change from opposite leaves to the $\frac{8}{21}$ arrangement in the bract-like sepals of the flower; but no distinction between bracts, sepals, and petals can really be made. *Chimonanthus*, however, would seem to be a more highly differentiated type, in that, not only is the calyx distinguishable from the corolla, but five exterior stamens constitute a distinct whorl by themselves, and the indefinite barren ones of *Calycanthus* are here reduced to five; so that, omitting the pistil, the flower consists of four distinct pentamerous whorls.

CHAPTER IV.

THE PRINCIPLE OF ARRANGEMENT.

SUPERPOSITION AND ALTERNATION OF WHORLS.—It has been already observed that leaves are arranged on two methods, either being on the same plane, *i.e.* opposite and verticillate; or with only one at a node, *i.e.* alternate. If the fibro-vascular cords passing from the leaves into the stem be traced downwards, those belonging to the leaves situate in one and the same vertical line always have their lower extremities inserted laterally and not actually confluent in that line, as will be seen in Fig. 7, taken from Hanstein's researches.*

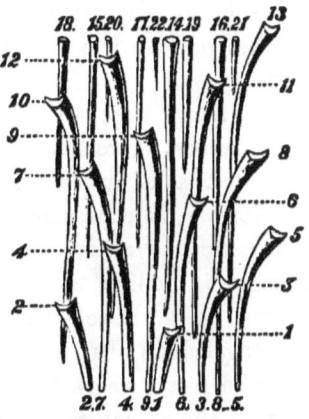

Fig. 7 —Diagram of the foliar cords in the stem of the *Arabis albida* (after Hanstein).

This fact is true, not only for foliage and bracts, but also to some extent for sepals and petals. When, however, we trace the origin of stamens and carpels, we find that their cords, instead of being inserted separately into the fibro-vascular cylinder, generally arise by branching, or by the so-called "chorisis"

* *De la Connexion qui existe entre la Disposition des Feuilles et la Structure de la Zone Ligneuse des Dicotylédons*, Ann. des. Sci. Nat., 4° sér., tom. 8.

of the cords belonging to the sepals or petals, or from both; and similarly the dorsal cords of the carpels branch off from the same stem as that of the sepals or petals, very rarely from both at once. Simultaneously with the dorsal, two marginal cords pass up directly into the placentas, having originated in the same way; and, in so doing, the floral receptacle usually becomes extinct, and takes, as a rule, no further part in the construction of the central portion of the pistil.

Starting, then, with these two fundamental sources of the various arrangements of the parts of flowers, we may first observe that of opposition or superposition and alternation, the former, if represented by decussate pairs of appendages, is the most primitive type. This is seen in many quaternary flowers in which the sepals emerge in successively decussating pairs. Such opposite leaves being foreshadowed in the cotyledons of exogens.

The next, or rather the first stage of differentiation is seen in the spiral condition which obtains in many flowers, mostly represented by the $\frac{1}{3}$ and $\frac{2}{5}$ types: thus, e.g., $\frac{1}{3}$ represents the arrangement prevailing in petaloid Monocotyledons; and all pentamerous calyces issue in a quincuncial manner. In *Sabia*, the petals follow continuously with the sepals in the same spiral line, so that the first petal is superposed to the first sepal. These whorls accordingly represent two cycles of the $\frac{2}{5}$ type, as seen above in *Garidella* (p. 21).

By far the commoner condition is to break up the spiral into cycles, say of five parts each, and then to shift their positions, so that they become alternate instead of superposed. Now, such a decussate arrangement is usually described as a fundamental law, not only governing opposite and verticillate leaves, but floral whorls as well; and particular stress is laid upon the usual presence of the petaline whorl of carpels,

THE PRINCIPLE OF ARRANGEMENT. 41

inasmuch as the law of alternation is thus carried out completely, and which may be represented as follows—the hyphens indicating the parts superposed to one another—Sepal-stamen; Petal-carpel.

From what has been stated above, the true order of arrangement and superposition would be — Sepal-stamen-carpel; Petal-stamen-carpel; and either one of the staminal and either one of the carpellary whorls may be suppressed. Thus, for example, *Oxalis*, *Zygophyllum*, *Geranium*, and *Ruta* have Sepal-stamen; Petal-stamen-carpel: while *Limnanthes*, *Coriarea*, and *Agrostemma* have Sepal-stamen-carpel; Petal-stamen. As instances where there is but one whorl of stamens, *Campanula* and *Hermannia* have Sepal-carpel; Petal-stamen; whereas *Linum* and *Diosma* have Sepal-stamen; Petal-carpel.

Of these variations, although Sepal-stamen is commoner than Petal-stamen, and Petal-carpel than Sepal-carpel,* yet these are, so to say, rather matters of accident than otherwise, in that it is probably due to certain exigencies of nutrition, and especially insect agencies, that such variations of arrangement exist.

The important fact mentioned above, that floral whorls are projected cycles and not primitive whorls, has, as far as I know, been entirely overlooked by botanists. Thus, for example, Professor Asa Gray remarks on the presence of whorls in flowers as follows: " Cycles alternating with each other are simply that of verticillate phyllotaxy," † to which he refers the opposite, ternate, quaternate and quinate verticils.‡ In the case of leaves, verticils represent usually more primitive types, such as twos and threes, and, from an evolutionary point of view, such *precede* alternate and spiral arrangements.

* *I.e.* in Exogens.   † *Bot. Text-Book*, p. 175.   ‡ *L.c.*, p. 120.

On the other hand, whorls of threes, and fives, and others in flowers are *compressed cycles* of spiral arrangements. They are, therefore, attempts at simulating ancestral or the verticillate conditions, but cannot possibly be primitive whorls themselves. That the petals can thus become decussating with the sepals is a result of the fact that their cords are not strictly superposed to and confluent with those of the latter. The total number of cords in the pedicel being usually limited to the same number as there are parts in the perianth, *i.e.* the calyx and corolla together, there is ample room for them to arrange themselves at equal angular distances around the central medulla of the pedicel. Then from the vascular cylinder thus formed, they pass off into the sepals and petals respectively.*

The sepals and petals or the two whorls of a perianth being thus provided for as to their fibro-vascular cords, the stamens and carpels, as already stated, generally depend upon these latter for their positions, and various arrangements arise according as the cords of the perianth-leaves give off new members or not. Theoretically there should be at least one whorl of stamens superposed to the sepals, another superposed to the petals, and two whorls of carpels as well; but while many flowers have both staminal whorls (*Caryophylleæ, Leguminosæ, Ericaceæ,* etc.), many others, as the *Gamopetalæ* retain only one, and more generally the first formed or sepaline, but sometimes it is the petaline, as in *Primulaceæ;* the probable cause in each case being certain exigencies in

---

* That foliar organs possess this power of rearranging themselves according to requirements is evident from other considerations; thus, many plants having freely growing erect shoots—as, for example, the common Laurel—have their leaf-arrangements represented by the fractions ⅔ or ⅜, but when extending horizontally, as in the usual condition, they are distichous. Similar features are seen in the Jerusalem Artichoke, which often changes its phyllotaxis on the same stem.

THE PRINCIPLE OF ARRANGEMENT. 43

the flower, through which nourishment is withdrawn at certain places to produce hypertrophy elsewhere. Thus the sepaline cord, instead of bearing an anther in *Primula*, bifurcates at the angle, and each branch proceeds up the margin of a lobe of the corolla, and aids in nourishing the latter.

As a converse instance of the sepaline cord undertaking a considerable amount of work, may be mentioned *Campanula medium*. In this plant the 5-lobed fibro-vascular cylinder of the pedicel sends off five cords intended for the calyx (Fig. 8, *sep.*); but, before reaching the base of the superior sepal, it sends off an innermost and lowest cord to become the dorsal one of the carpel (*d. car.*), which, in this flower, is thus superposed to a sepal. It also sends off two, right and left, one for each petal alternating with it (*pet.*); so that each petal receives two cords, one from each adjacent sepal,—a most unusual condition of things, for petals have almost invariably their own cords issuing from the pedicel. Lastly, the same sepaline cord provides that of the stamen (*st.*) superposed to it. In this flower, therefore, we can understand why there is no petaline whorl of stamens; simply because the corolla does not possess its own proper fibro-vascular cords to give rise to them.

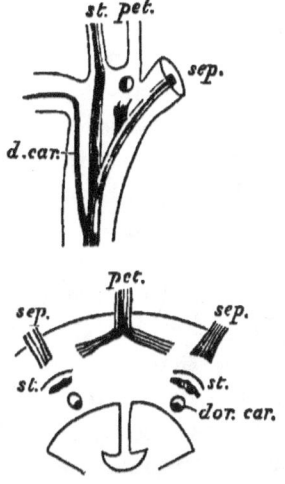

Fig. 8.—Vertical and transverse sections of the wall of the inferior ovary of *Campanula medium* (after Van Tieghem).

On the other hand, in the *Malvaceæ* after the axis has supplied cords for the sepals, others furnish those of the corolla; these latter, however, by *radial* division form two to each petal, subsequently dividing into several; for the same pair by repeated *tangential* division gives rise to the series of stamens (which have been thus doubled) superposed to each petal, both having arisen from a common cord.

With regard to the numerous carpels of Hollyhock, I find that the axial cylinder which has given rise to the five sepals continues on, and by radial division again supplies cords to the carpels, which are grouped into five sets superposed to the sepals, as may be easily seen if the pistil be examined from below. Hence, as the sepaline or petaline cords in these flowers each undertake to form a large number of extra parts—many stamens in the one case, and many carpels in the other—it is presumable that neither sepaline stamens nor petaline carpels could be formed.

With regard to the presence, and consequently the relative position, of one whorl rather than the other of the gynœcium, it is due to the fact that sometimes the sepaline cord will give rise to the dorsal carpellary, as in *Althœa* and *Campanula*; at others, it is the petaline, as in *Fuchsia*, *Sedum*, *Ivy*, etc.; so that the carpels become superposed to the sepals or petals accordingly. As instructive instances of variations in this respect occurring in the same family, it may be mentioned that all species of *Campanula* which have five carpels, as also *Wahlenbergia capensis*, *Michauxia*, *Canarina*, and *Lightfootia subulata*, have their carpels superposed to the sepals and stamens. On the other hand, *Musschia* (*Campanula aurea*, L.) *Platycodon* (*C. grandiflora*, Jacq.), and *Microcodon* have the carpels superposed to the petals.

The fact that either the sepals or the petals can have the carpels superposed to them respectively, just as they can each have a whorl of stamens, and that, in some few orders, the two whorls are actually present, as in *Butomeæ* and *Juncagineæ*, led me to assume two whorls as the primary

THE PRINCIPLE OF ARRANGEMENT. 45

or ancestral number of carpels in an ideally complete flower.

Besides the usual alternation of whorls resulting from a regular and equal displacement of every part of the whorl, there may be unequal displacements; thus, while *Cistus* has a pentamerous flower, with strict alternation of its whorls, *Helianthemum* has a tendency to be trimerous; first, in the two outer sepals being reduced in size, and the pistil to three carpels instead of five. In this flower there are five petals, but in correlation with the preceding irregularities, it will be found that two pairs of petals stand superposed to the sepals, Nos. 3 and 5, while a single petal is over No. 4; Nos. 1 and 2, therefore, have none superposed to them. With regard to the stamens, it may be added that those of *Cistus* consist, first, of one whorl of five, the most interior and first developed superposed to the sepals; and a second whorl superposed to the petals, in which the stamens are grouped into five clusters. The staminal whorls arise centrifugally.

Another cause of a change of order in the whorls results from substitution of one kind for another. Thus, in the female flower of *Zanthoxylon*, the five carpels are superposed to the five sepals. In the male, five stamens now occupy exactly the same place as the carpels, the corolla alternating with the sepals in both kinds.*

The interpretation I would suggest is that the sepals, being the only whorl of the perianth developed, the calyx is the only source for supplying the dorsal cords of the carpels which thus become necessarily superposed to them.

From what has now been said, it will be seen that the arrangement of the essential organs of a flower is, as a general

---

* See Figs. in Le Maout and Decaisne's *Descriptive and Analytical Botany*, p. 324. The female flower is described as apetalous, but Payer discovered rudiments of the petals.

rule, most intimately connected with the union of their fibro-vascular cords with those of the perianth; and as parts of flowers are often multiplied, as the petals of *Camellia*, perianth-leaves of Daffodils, etc., such has given rise to the idea of chorisis or *dédoublement* of French authors; as if one organ had split into two or more. That vascular cords can become repeatedly bifurcated is abundantly observable, whether radially, as in the case of the carpels of the Hollyhock, or tangentially, as in producing the stamens of the same flower. The more correct way, therefore, of regarding the process would seem to be, first, to recognize the phyllotactical origin of the perianth as the basis to start from, and then to regard each fibro-vascular cord as an instrument for furnishing any number of appendages, whether they be additional petals, stamens, or carpels, by the process of chorisis, not of the complete organ, as generally meant, but of the cord belonging to it.

To summarize these remarks—we find that the cause of the alternation of the whorls of the perianth, or of the calyx and corolla, is due to their being made up of cycles of spiral arrangements, which are projected on to the same plane, and so form verticils. Their positions are then shifted so that the parts of each whorl bisect the angles between the parts of the whorl succeeding or preceding it.

Secondly, having laid this foundation, the stamens and carpels follow in superposition to one or other or both of the preceding whorls in consequence of the branching of the fibro-vascular cords. And this accounts for superposition.

It may be still further inquired why in some cases the sepaline, and why in others it is the petaline cords which give rise to a whorl of stamens or carpels, as the case may be. The reply at present must be speculative, for there may

THE PRINCIPLE OF ARRANGEMENT. 47

be more than one influence at work to determine what whorl shall follow each of those of the perianth.

The *immediate* cause is nutrition; but the deeper question, what directs the nutrition to one cord rather than another, can only be guessed at in most cases: but as the petaline stamens are generally absent from at least the *gamopetalæ*, it would seem that the enhancement of the corolla through the agency of insects has caused the whorl of stamens in front of it to be atrophied through compensation. Some special circumstance, however, we know not what, have interfered to retain that whorl in *Primulaceæ*, and some few other plants.

The reader must be reminded, however, that this method of branching in order to give rise to stamens and carpels from the cords of the perianth is not universal. When they are many, it is done by the fibro-vascular cylinder of the pedicel becoming much enlarged, and consisting of a great number of cords, all arising by lateral chorisis, it is true, but long before they enter the floral members; so that by the time the latter are about to emerge they each receive their own cords from the general axial cylinder. This is what happens *e.g.*, in *Ranunculaceæ* and *Cruciferæ*.

CHAPTER V.

THE PRINCIPLE OF COHESION.

COHESION.—GENERAL OBSERVATIONS. This term signifies the union between parts of the same kind or whorl; and the prefix gamo- is used in conjunction with the terminations -sepalous, -petalous, and -phyllous,—to indicate that the parts of the calyx, corolla, and perianth respectively cohere. In the case of the stamens, they are said to be mon-, di-, tri-, or poly-adelphous, according as the filaments cohere into one, two, three, or more groups; while syngenesious is used for the coherence of anthers, and, lastly, syncarpous denotes that the carpels of a pistil cohere.

There are two kinds of cohesion, congenital and by contact.* Congenital cohesion I regard as an advance upon freedom, or a further state of differentiation; for, according to the principles of Evolution, freedom or separation of parts must precede their union; just as, for example, bones are free in the embryo which become "ankylosed" in the adult; or always free in a fish, while their homologues cohere in higher types of vertebrates.

Congenital cohesion applies to by far the greater number of cases of union amongst the parts of the different whorls

---

* We might appropriately distinguish these two kinds of union by the terms *connate* or "born together," and *coherent* or "sticking together."

of flowers, respectively. Cohesion by contact is the cause of the anthers being syngenesious in the *Compositæ*. It applies, sometimes at least, to the two margins of each carpel when in contact up the axis of an ovary, as of that of a Lily. The stigmas of Asclepias are at first free, but later in their development they become coherent by contact.

Congenital cohesion takes place almost from the very commencement of growth and development of the parts, so that when full-grown there may be no trace of the line of cohesion. Fibro-vascular cords, indeed, often occur in the very position of it, not unfrequently branching off in various ways, as, *e.g.*, at the fork to nourish the adjacent free portions of the limb. This occurs in the calyx of *Stachys* and the corolla of *Primula*, etc. In *Campanula rotundifolia* the fibro-vascular system of the corolla becomes completely altered, and instead of representing that of distinct leaves in contact by their edges, the veins ramify and anastomose all over the general space between the two adjacent dorsal ribs, completely obliterating all trace of the line of union between them. In the case of the Primrose, however, the calyx has the exact appearance of five pinnately nerved leaves being united by their thin and impoverished edges, where there is nothing but translucent tissue without any cords at all.

It is important to observe this more or less complete modification of the fibro-vascular system under congenital cohesion, as it shows how much more highly differentiated a condition has been acquired than when the parts are free. In the latter case they represent more closely the forms and venation of distinct foliar organs.

As a curious instance of cohesion of both kinds in the same organ, may be mentioned the corolla of *Phyteuma*; the basal portion of which consists of five petals congenitally united; but the five portions of the limb cohere by contact at the apex, and so form a tube which collects the pollen shed into it by the five free anthers, which are included within this corolla-tube (Fig. 9). They thus form the "cylinder" for the "piston" action of the pistil which continues to grow, and so sweeps out the pollen beyond the extremity of the tube, just as it does from the syngenesious anthers of the *Compositæ* and *Lobelia*. The five portions of the corolla thus cohering by contact subsequently become more or less free.

Fig. 9.—*Phyteuma* (after Müller).

The rationale of Cohesion lies in its adaptation to insect agency, and implies a greater degree of specialization than when the parts of the whorls are free. Thus in *Thalamifloræ*, of such an order as *Ranunculaceæ* with regular flowers and with all the parts of the perianth whorls free, the flowers are usually visited by a much greater number and variety of insects than are those of orders of *Corollifloræ*. For example, Müller records sixty-two species of insects as seen by him to visit *Ranunculus acris;* whereas the humble-bee alone enters the gamopetalous tube of the Foxglove. This adaptation of *form* to insect visitors will be better appreciated when we come to discuss that principle of Variation, which so powerfully affects floral structure.

It occasionally happens that parts normally united become free: the process is called "dialysis," and may be regarded as a reversion to an ancestral free condition. Fig. 10 represents a flower of *Mimulus* in this condition. The rationale of cohesion in the sepals, petals, and stamens, I regard as the immediate result of hypertrophy set up by insect agency,

THE PRINCIPLE OF COHESION.

aided by the close proximity of the parts; and as a resulting effect, is the ever-increasing adaptation to the requirements of insects, which are more and more specialized for them, so that, for example, Lepidoptera are almost solely adapted to long tubular flowers like the Honeysuckle.

Fig. 10.—*Mimulus* undergoing "Dialysis" (after Baillon).

An analogous process of congenital cohesion is well seen in the fasciation of stems which occurs particularly often in succulent shoots, as Asparagus, Cabbage, Lettuce, and the young shoots of the Ash tree. This is most reasonably referred to hypertrophy coupled with the close proximity of the buds which ought to have developed into independent shoots. Again, cohesion between the sepals or petals of Orchids is not uncommon abnormally under cultivation; and would also seem to be due to the stimulating conditions under which they are artificially cultivated.

Hypertrophy in an organ is due to a special flow of nutriment to it; and cohesion may result from the close proximity of the parts of the whorl to one another; but the influence which brings about the determination of sap to a particular point, I take to be the mechanical strains induced by the insect visitors when alighting upon the flower in search for nectar or pollen.

If this principle be correct, that the tubular structure of calyces and corollas, as we see them now, has arisen through the requirements of those organs to meet strains thrown upon them; I think it will furnish the solution to many a question that may arise as to the peculiar shapes of corollas, etc., besides explaining the very principle of cohesion itself. An insect alights on one or two petals. In order to support it, an immense gain is secured if the flower can call in the aid of the other petals; and this is obviously obtained by their cohesion into a tube, just so far as the required strength is wanted. Nothing would be gained by the portions of the limb being united, as far as additional strength was required to bear the burden. The tubular structure is the strongest possible, and when short, as in rotate corollas, little extra aid is required; but if it be long and visited by heavy insects and not by Lepidoptera, which hover in front of the flower and only insert their long and slender proboscides, then the tube finds additional support in the calyx being tubular as well. At other times mutual support is gained by the close contact of the flowers, as in a capitulum of the *Compositæ*, from which the calyx vanishes.

Of course, every degree conceivable is met with between short, stout, and strong tubes with no additional aid, and slender ones supported by a strengthened gamosepalous calyx. These are adapted to insects which alight upon the corolla limb; while for Lepidoptera the tube is more elongated, and, as no weight is thrown on the anterior petals, no extra support is required. That this is the true interpretation of the origin of a gamopetalous corolla, appears from such negative evidence as is seen, for example, in *Lonicera Periclymenum* and *Asperula taurina*,* which have greatly elongated and contracted tubes, deriving no support from the arrested calyx; and although somewhat two-lipped, the anterior member is no larger than the others; the reverse being always the case when a heavy insect is the regular visitor. These two species are exclusively fertilised by the Lepidoptera, such as the Hawk-moth, which only hovers in front of the orifice, but throws no weight upon the corolla.

* See Müller's figures, *Fertilisation*, etc., pp. 296, 303.

We may see, as it were, Nature's first attempt to form a tubular process in the *Cruciferæ*. Here it is obtained by simple approximation of the slender claws of the petals, which are supported by the erect and closely imbricated sepals. A step further is gained in Dianthus, in which the sepals cohere but the petals are still free. The third and last stage is arrived at when both calyx and corolla are tubular.

Subsequent to this state of cohesion many additional structures may arise as they are required in the formation of ribs, etc., as already explained; while the very form of the tube may change from a purely straight cylinder to a curved or expanded funnel, etc., according as special strains have to be met, which the original form was not well calculated to sustain.

These changes of Form will be more fully discussed when I treat of that principle of Variation.

CHAPTER VI.

THE PRINCIPLE OF COHESION—*Continued*.

COHESION OF THE SEPALS, OR GAMOSEPALOUS CALYX.—This is congenital, and may be free, as in the Carnation and Primrose, or associated with a "receptacular tube," as in *Leguminosæ* and *Rosaceæ*.

As sepals mostly represent the petioles of leaves, the tubular part of a gamosepalous calyx consists really of the fusion of the expanded petioles, the teeth of the limb being all that remains to represent the blades which are usually suppressed. The main fibro-vascular cords correspond to the mid-ribs, while the interspaces are either without additional "marginal" cords, as in the Primrose, or with single or double cords in the line of junction, as in the *Labiatæ*; or they may be covered with anastomozing reticulations without any linear cord at all, as in *Mimulus*.

With regard to the presence of linear cords in the line of suture, if there be five sepals, there will be at least ten ribs to the calyx; *i.e.*, if there be only one marginal cord; but as there are two margins which cohere, they may have a separate cord apiece; and then there may result fifteen cords in all. Thus *Stachys* has five dorsal cords with barely traces of five marginal ones; *Ballota* has ten, and *Nepeta* fifteen.

The above arrangements may be modified by the separation of the two marginal cords in certain places but not in

THE PRINCIPLE OF COHESION. 55

others, while supernumerary cords can be formed, which appear to have for their function to strengthen the calyx to meet the strain upon it when an insect alights upon the flower.

In the calyx of some species of *Salvia*, which is strongly bi-lobed, though retaining its five teeth, three dorsal (d) are posterior and two are anterior. There are two single marginal (m) cords between the three posterior and dorsal, which correspond to the mid-ribs of three sepals. The two lateral and marginal cords are each double; while a supernumerary cord (s) lies beneath the lip of the corolla between the two anterior marginals. The accompanying diagram of the sepaline cords of *S. Verbenaca* will illustrate the arrangement.

The arrangement of the cords (m and s) shows that the strain being greater on the anterior side, the calyx has, as it were, *stretched* in that direction, the two marginals having separated so widely in front, as to require an extra cord (s). The two lateral ones have not separated to so great an extent, while on the posterior side, where little or no strain is felt, the marginal cords have remained single.

As the cord (s) shows how Nature can add a fibro-vascular cord if required, so one or more can be subtracted by atrophy where no stress occurs. Thus the petals of the *Compositæ* have no dorsal or median cords, the five sepaline only being present below, but pass up the margins of the petals. Conversely, in the Primrose, the calyx, giving no support to the corolla, has no marginal cords.

The above diagram will represent the distribution of the sepaline cords of *S. glutinosa* and other species, as well as *S. Verbenaca*, but in *S. pratensis* the strain has apparently been not so great, consequently the supernumerary cord (*s*) has not been developed.

Such slight differences are significant, because they show how readily an organ can respond to different degrees of force brought to bear upon it by different insect visitors; and the cords are invariably placed just where the strains are greatest.

The number of ribs to the calyx has been adopted by systematists as generic characters in some of the *Labiatæ*, as well as the tubular or campanulate shape of it. Now, it will be found that the shape corresponds with the requirements of the corolla; so that if the tube of the latter be comparatively short and slender, the calyx completely encloses it, and has its surface strengthened by a variable number of ribs according to the genus; though they are not always constant on the same plant. As examples, may be mentioned, *Mentha* and *Melittis*, which have a broad campanulate calyx, and a broad tube to the corolla. *Stachys* has 5-10 ribs surrounding the cylindrical corolla-tube. *Galeopsis versicolor* has 10 prominent ribs, and 10 others which reach from the base of the calyx-tube to about half-way up. *Melissa* has a very narrow elongated calyx, which fits the slender tube of the corolla exactly, and has 13 or 14 ribs.* Similarly *Nepeta Cataria* and *N. Glechoma* support the contracted slender basal part of the corolla-tube, and have 15 ribs to the calyx.

Teucrium Scorodonia has only 5 dorsal ribs and 2 (posterior) marginal. The calyx is very broad compared with the slender corolla-tube, and scarcely, if at all, supports it. This flower, is visited both by bees, and nocturnal Lepidoptera which suck without throwing any weight upon the flower.

COHESION OF PETALS, OR GAMOPETALOUS COROLLA.— As

* This difference in the number of ribs depends upon the lateral and marginal being single or double.

already stated, this is congenital, and, as with the calyx, so with the corolla, the line of junction may be marked by a marginal cord, or the interspace covered with reticulations as in *Campanula rotundifolia*.

As in the calyx of many Labiates, so there may be supernumerary cords in the corolla, until they may be greatly increased in number, as in *Convolvulus Sepium, Digitalis*, etc. The cords being straight in the tube may ramify in the lobes, adding thereby marginal veins to the latter, as in *Primula* and the *Compositæ*. In this last, the petals are devoid of median nerves, hence the importance of the marginal with their branches up the edges of the corolline lobes.

It would be superfluous to multiply examples if the principle be understood; and what I particularly wish the reader to realize is the, so to say, extraordinary plasticity which resides in these organs of flowers, in that they evidently have the power of altering their structure to meet a variety of requirements; so that if we might compare them to architectural buildings, we might say that the floral Architect at one time saw not only a chance of some ornamental improvements in a frieze at some particular place, graceful lines of colour or curvature in another; or, again, flutings, depressions, and elevations, etc., all breaking up any chance of monotony: but cunningly adds elegant buttresses without, as well as runs up ribs of masonry within the walls; which, while intended to meet particular strains, only add additional charms to the general and harmonious beauty of the entire fabric.

COHESION OF STAMENS—(1) "ADELPHOUS" FILAMENTS.— This occurs in various degrees, from a comparatively slight union at the base, as in *Linum usitatissimum*, to a short distance from the anthers, as in *Malvaceæ* and *Leguminosæ*. It is undoubtedly an adaptation to insect agency.

If the stamens be monadelphous, and the union be extended, it may completely enclose the usual honey-secreting surface characteristic of allied genera, the result being that it can secrete none at all. In such cases, insects are deceived in visiting the flower, as in *Genista,* and some other monadelphous genera of *Leguminosæ.* Otherwise, the honey is secreted by some other source external to the staminal tube, as in *Linum catharticum;* in which flower five inconspicuous glands occur on a fleshy ring, just opposite the stamens. In *Malva,* the honey is found in five pits between the bases of the petals, and in *Pelargonium* in a long tube formed by one sepal, the insertion of which remains far below that of the others, which are carried up by the growth of the pedicel. In *Laburnum,** as in *Orchis,* instead of a secretion, the fluid is only to be secured by piercing succulent tissue which is found in front of the vexillum in the form of a cellular cushion.

In diadelphous species of the *Leguminosæ,* the honey may be secreted by the inner basal portion of the staminal tube,* or else, and perhaps more usually, by an annular disk which surrounds the short pedicel of the ovary, as in *Pisum.* In this case the honey is easily secured by the proper insects as the superior stamen is free, and there is also an additional facility of access by means of an oval space formed by the widening of the staminal tube just above their base.

In *Cercis,* the disk is very large, and the 10 stamens stand in depressions around it. Consequently they are entirely free.

The staminal tube, together with the petals, which are more or less interlocked together, protect the honey from being rifled by the wrong insects,† as it can only be secured

* According to Müller.
† A curious additional protection occurs in *Hippocrepis comosa,* in that the claw of the vexillum, which is elevated in a remarkable manner,

by such as have proboscides of sufficient length to reach it, corresponding, of course, to each species or genus.

Papilionaceous flowers being irregular, and visited in but one way, it is only the superior stamen which is free; but the staminal tube is often imitated in other flowers where there may be no cohesion at all, as by the tribe *Ocimoideæ* of Labiates, *Collinsia bicolor* of the *Scrophularineæ* and *Polygala*, etc. Similarly, in the case of regular flowers, the monadelphous condition may be closely mimicked by filaments which are stout and sufficiently rigid to form a column. This occurs in *Cruciferæ*, *Viola*, *Convolvulus*, *Crocus*, etc. In some cases, as in *Crambe* and *Deutzia*, the filaments are provided with wing-like structures which render the tube more complete. In orange flowers, a certain amount of cohesion is actually obtained between some of the filaments.

(2) SYNGENESIOUS ANTHERS.—These, as stated, are not congenitally united, but by simple contact. As with filaments, so with these, it is an adaptation to insect fertilisation. *Jasione montana* furnishes a good instance for an incipient stage where they just unite at their bases only. This cohesion is completed in the genus *Synanthera* of the same order *Campanulaceæ*, as well as in the sub-order *Lobelieæ*. In other cases of true syngenesious anthers there is a complete lateral fusion, as in *Lobelia* and *Compositæ*, in *Gloxinia* and *Impatiens*. In all these cases the cohesion is by lateral contact only, and not congenital; that is to say, the papillæ of the future anthers on emerging from the axis grow to a somewhat considerable stage of development as incipient anthers before coming into contact. They then coalesce, apparently by a slight solution of the surface of the cellular

carries a triangular flap, which exactly covers the orifice leading to the honey. A somewhat similar flap occurs in the petals of *Phaseolus* and *Delphinium*, which likewise keeps out unwelcome guests.

walls which touch; so that when they are fully grown the cohesion is firmly secured. An imitative cohesion is seen in the anthers of the Heartsease, which arises from the interlocking of marginal hairs down the sides of the cells. Anthers, when thus closely approximate without actual cohesion, are usually called "connivent," as in *Ericaceæ*, and the word is perhaps appropriate to those of *Solanum Dulcamara*; but in this plant the union is very close, and might even be considered as syngenesious.

The rationale of the close approximation of anthers, or of actual cohesion between them, is the effect of insect agency, just as for the filaments; but the method of extraction of the pollen varies. In *Viola*, the proboscis is thrust through a small orifice between the connectival appendages of the lower pair of stamens, in order to reach the end of the honey-collecting spur. In Heaths and some of their allies, the anther-cells are at first in contact, and so prevent the pollen from escaping; but each anther is provided with two auricles which extend to the corolla. A bee on entering first strikes the projecting stigma, but its proboscis soon turns one of the auricles aside, which, acting as a lever, dislocates the rest, and a shower of pollen falls out.

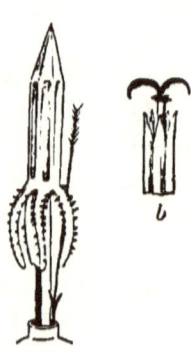

Fig. 11.—Stamens of *Centaurea*.

In *Compositæ* and *Lobelia* there is a true piston action. The style continuing to elongate drives the pollen out of the cylinder formed by the anthers, and elevates it above the flower, thereby rendering it easy to be dispersed by insects. This is well seen in *Centaurea* (Fig. 11); (*a*) represents the stamens with the anther-cells closed above by the connectival appendages. The arrow shows the direction of the insertion of the proboscis of a bee to reach the annular honey-disk at the base of the style; in *b*, the style-arms have spread after protrusion through the separated connectives. The brush-like tuft of hairs has swept the pollen out by means of the piston-action of the style.

In *Campanula*, the action is different, for the anthers though connivent, have not yet become syngenesious, as in allied genera, *e.g. Lobelia*. They at first closely surround the style, which is provided with long collecting hairs upon which the pollen is caught. The anthers then shrivel and fall down. Subsequently a bee enters the expanded bell, grasps the style with her legs, and so transfers the pollen to the abdomen. This method is identical with that followed by bees in getting honey from *Crocus*, though in this genus the anthers remain erect, and, being extrorse, at once discharge the pollen upon the insect without the intervention of the style.

CHAPTER VII.

THE PRINCIPLE OF COHESION—*Continued.*

COHESION OF CARPELS, OR SYNCARPOUS PISTIL.—The accepted doctrine that the carpels are metamorphosed leaves, will be considered more fully when teratological modifications come to be discussed; and the proof that an ordinary carpel, such as a legume, is merely a leaf folded upon itself in a conduplicate manner with the margins coalescing and then metamorphosed into a new organ, requires no special evidence now. That a syncarpous pistil consists of two or more carpellary leaves coalescing is equally admitted; and there are two methods of cohesion. Either the carpels may be *ab initio* composed of unclosed leaves, which cohere by their edges * respectively in contact, thus forming a single cavity provided with parietal placentas,—such a union implying a more primitive or arrested condition, from an evolutionary point of view; † or they may be individually more or less closed before coalescence takes place, in this case by their lateral surfaces. The axile placentation is the result. The

---

* The theory that the placentas are, at least in part, axial, will be seen to be erroneous in consequence of the orientation of their vascular cords (*e.g.* Fig. 12, *c*, p. 64; and Fig. 13, *a*, *b*, p. 65).

† Thus the parietal placentation of *Orobanche* is probably a result of degradation through parasitism, from the axile, of the *Scrophularineæ*. It may be compared to a "cleft palate" and "hare-lip" in man.

THE PRINCIPLE OF COHESION.

margins show every degree of union from a mere contact without real cohesion, thence, cohesion by contact, to a solid central axial structure formed by congenital cohesion. Lastly, the ovary may be one-chambered, with a free-central placenta, as in *Caryophylleæ* and *Primulaceæ;* or with one or more ovules attached at the base, as in *Rumex, Compositæ, Gramineæ,* etc. It is these latter kinds especially which have given rise to much discussion as to the real nature of the placentas, and as to how far the axis enters into their construction. To ascertain this latter point, a study of the distribution and structure of the fibro-vascular cords of the axis and of the carpels would seem to afford the most promising clue to the interpretation.

It has been already mentioned that the dorsal cords of carpels generally arise by lateral division from those of the sepals or petals; and then the carpels will be superposed to the one or the other of these organs respectively; * or, a group may emerge from the axial cylinder in a horse-shoe form, as seen in section; the outermost cord becoming the dorsal-carpellary, and the ends of the curve the marginal. This is the case, for example, in *Cyclamen.*

The point, then, at which the carpellary cords branch off from a common stem in the first case may be regarded as marking the termination of their axial character; and in the latter case, at the separation of parts of the "horse-shoes" to form groups of threes. With regard to those cords which become marginal and placentary, it is important to notice the position of their spiral vessels.† If they are situated on the side of the cord nearest to the medulla, the cord may

---

* See pp. 23, 24, and 42, 43, 44.

† The cords are, of course, reduced to vessels and soft bast only, the former being mostly spiral, but occasionally becoming more or less reticulated. I shall adopt the usual word *Tracheæ.*

generally be regarded as axial; if, on the other side, *i.e.* nearest to the ovary-cell, and if transverse sections exhibit intermediate positions, in which they are central or scattered irregularly within the phloëm, they are then marginal and placentary.

They may change their position from one side to the other of the cord, as far as I have observed, in three different ways. The whole cord may twist to the right or left, as in Hellebore (Fig. 12); or, secondly, it may divide into two, and each half turn towards an adjacent half of another cord and unite with the latter, as in *Pelargonium zonale* (Fig. 13, *b*); or, thirdly, the tracheæ may traverse the phloëm and so pass out at the opposite side at a higher level, as in Ivy (Fig. 14, *f*, p. 68). In any case, as soon as the tracheæ are so placed as to effect their object of nourishing the ovules, they may be pronounced to be unquestionably and strictly carpellary.

I will take Hellebore as illustrating the first case. Fig. 12 represents a section of the floral receptacle taken imme-

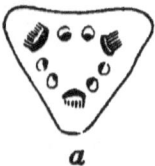
a

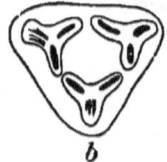
b

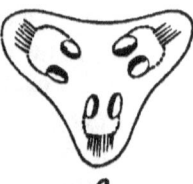
c

Fig. 12.—Hellebore: sections at base of ovary.

diately above the insertion of the innermost stamens. There are nine cords* oriented as axial, three of which are beginning to curve outwards to form the dorsal cords of the three carpels. Sections made a little higher show that the three pairs of cords have spread out and revolved so as to bring their spiral vessels into a radial direction (*b*, *c*). In this

* The tracheæ are indicated by black lines or dots, the phloëm being inclosed within the thin lines.

THE PRINCIPLE OF COHESION.

position the tracheæ of each pair of cords face each other. At this point, then, they have quite lost their strictly axial character of facing the centre, and the axis is therefore no longer concerned in the structure. A little higher the cavities of the ovaries (indicated by the dotted lines) appear between the dorsal cord and the pair of marginal ones; and now the latter turn their spirals completely towards the ovary cells, having rotated through 90° in all. The object of this rotation is to enable them to send off cords to the ovules.

The second method is well seen in *Geranium, Pelargonium zonale*, and *Impatiens*. A section of the receptacle of the first two, made between the insertion of the stamens and the pistil, shows five groups of three cords each, arranged as in Fig. 13, *a*. Small portions of the ten staminal cords are

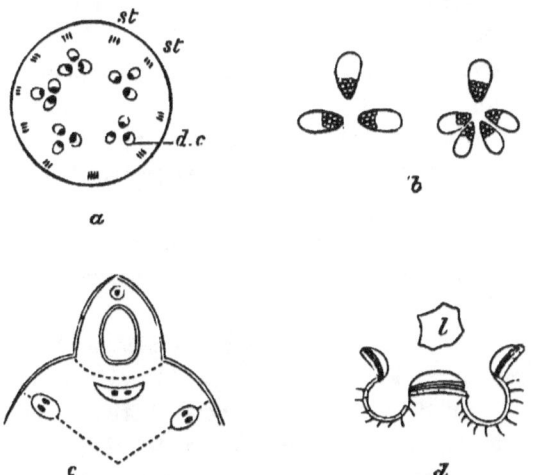

Fig. 13.—*Pelargonium*: sections at base of ovary (*a*, *b*, after Van Tieghem).

seen on the circumference of the section. The outermost one of each group of three will form the dorsal cord of the carpel. The two inner have their vessels already turned towards each other, as described in Hellebore, and are in part required for the placentas. They are, therefore, no longer oriented as in an axis, *i.e.* with all the vessels arranged on the inner edge of the cord and facing the central medulla.

A short distance above the base of the pistil, the innermost cords divide in a somewhat irregular manner, but rearrange themselves symmetrically round the centre of the ground tissue in ten cords, as soon as the ovary cells have put in an appearance. The method by which this condition is arrived at was described by Van Tieghem in *Geranium longipes*, and with slight modifications it will apply to *Pelargonium zonale.* Each of the lateral cords divides into two (Fig. 13, *b*), the two interior and adjacent branches unite to form a single marginal cord with the tracheæ within or on the outer side (Fig. 13, *c*). The two outermost branches pass off to the right and left, and proceed to join the corresponding halves from the neighbouring systems. The pairs uniting thus form five cords of double origin, alternating with the crescent-shaped marginal cords of the carpels (*c*). There are thus formed five in front of the ovary-cells, and five in front of the septa; "which," Van Tieghem observes, "one would regard as axial, if one did not pay attention to the mode of formation of the cords and to their orientation."

In his description of *Impatiens Royleana*, he says that the two innermost branches (Fig. 13, *b*) unite at first end to end, *i.e.* like an 8, with the tracheæ at the extremities in contact; they then form one cord with the spiral vessels towards the circumference of the section, by rotating through 90°, accompanied by complete fusion.

In *Pelargonium zonale*, the tracheæ become plunged, as it were, within the phloëm-tissue of the cords, as shown in Fig 13, *c*, which then fuse together laterally.

Above the ovary-cells, at the base and thicker part of the style, a section (Fig. 13, d) shows five solid circular buttresses, the tissue of which is continuous with the central parenchyma, in the middle of which a lacuna (l) is formed by rupture. In the depression between the buttresses, a small portion of the style and conducting tissue forms a bridge, as in Fig. 13, d, showing a cavity below it.

It is in this homogeneous mass of ground tissue that we have a complete fusion of the hypertrophied borders of the carpels which have thus entirely lost their individuality. The axis proper disappeared as soon as the spiral vessels became oriented, as in Fig. 13, a.

Hence the dotted lines radiating from the centre (c) mark the ideal boundary of each carpel, and the line across the base of the ovary-cell is the place where rupture will take place when the fruit is mature. The column, or so-called "carpophore," remaining is therefore entirely carpellary in its origin.

The third method by which the tracheæ pass from one side to the other of a cord is partly seen in the preceding; and I suspect that this is the commonest method of all; for though, when axial, the cord has its spiral vessels fixed at the inner angle, as soon as a change of position occurs or whenever it has to branch, the fixity of the position of the tracheæ becomes relaxed, and they readily become enveloped in the rest of the tissue of the cord, and so pass from one side to the other with perfect facility, as will be seen in the case of the Ivy.

When a syncarpous pistil has its ovary inferior—that is, imbedded in the receptacular tube—the real state of cohesion between the several carpels is masked in consequence of their partially undifferentiated state; the ovaries of which then have the appearance of being simply isolated cavities sunk within a mass of parenchymatous tissue. In fact, they might often be called "falsely syncarpous," a term applied to the *Pomeæ*, but which is equally applicable to Ivy and Fuchsia.

In the pedicel of a flower of Ivy, there are, at a distance of about three-quarters of an inch from the tapering base of the inferior ovary, four fibro-vascular cords (Fig. 14, *a*). A little higher these split up into an irregular circle (*b*), and shortly above the base of the receptacular tube there are fifteen (*c*), ten being more towards the circumference than the other five. The outer ten are for the sepals and petals. The five inner will appear superposed to the sepals (*d*), having been already separated off by radial chorisis rather low down; these are for the stamens. Then from the petal-ine cords, by a similar method of chorisis, a small cord runs up the dorsal part of the ovary-cell and another up the axis. This fixes the position of the five carpels (if so many be present) as superposed to the petals (*d*). There are often only four, or even three,

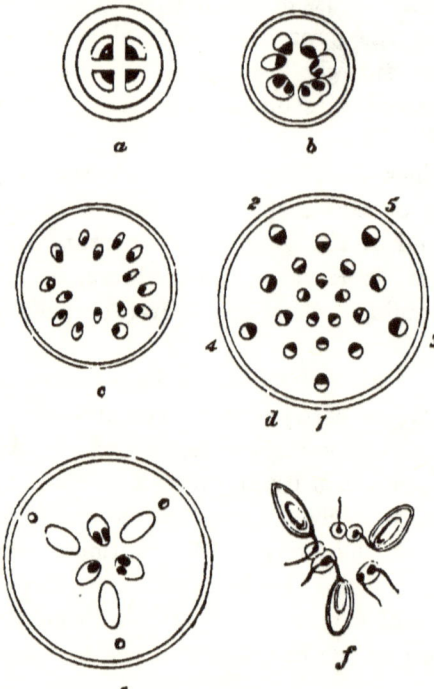

Fig. 14.—Ivy: sections from pedicel to summit of ovary.

THE PRINCIPLE OF COHESION. 69

ovary-cells developed. When this is the case, the cords of the centre become fused into four or three $(2 + 2 + 1)$ (*e*), and take up a position *alternating* with the ovary-cells. They become even more welded together higher up; but they separate again, to form twice as many as there are ovary-cells (*f*). If there be three, then each cord may bifurcate, though they do not all do so in every instance; so that out of 12 cords, three ovular cords are given off to nourish the ovules (*f*), and the rest run up the styles, though the total number of cords may be less than 12, as variations seem to take place.

The ground tissue consists of a loose merenchyma, excepting three or four layers of cells below the epidermis, which are more compact; the ovary-cells—seemingly reduced to a thickened epidermal layer only—are plunged freely into this tissue (*e*). The cords run up the centre perfectly independent of the ovary-cells (*e*) with their spiral vessels on the inside, surrounding a central medulla. Were it not for the presence of the dorsal cord, there is nothing to hinder one from calling them axial. It is not until they reach the top of the ovary-cells that these cords bifurcate and send off one branch each into the pendulous ovules, the other branches being conveyed upwards into the styles (*f*).

The above description will give a fair example of the distribution of the cords for supplying the several members of the whorls. The reader can estimate how far the central cylinder should be called axial. The fact is, that the whole of the tissue of the carpels, excepting the thickened internal epidermis covering the ovules, is totally lost in the general spongy mass in which they are imbedded. But since the petaline cord gives rise to the small dorsal-carpellary and one axial, theoretically these two belong to the carpellary leaf; and on *this ground* we should feel inclined to regard the central cords not as axial but marginal and carpellary, notwithstanding the fact that the tracheæ are oriented inwards; since it is not until they reach the level of the insertion of the ovules that they pass either to the middle or opposite side of the cord. The rest of the carpellary tissues are undifferentiated, as stated above, and it is this very common condition in the case of inferior ovaries that has led botanists to regard the lower parts of the carpels as being of an axial nature and not foliar.

THE FORMATION OF SEPTA.—With regard to the union of the surfaces of the carpels to form the septa, the rule is for the adjacent epidermides to be altogether wanting; and, if the median tissue be thick, the walls of two adjacent ovary-cells may be very wide asunder, as in the Ivy. On the other hand, the septa may be reduced to the two epidermal layers alone, and then they are often scarcely coherent at all, as in Balsam and Lemon.

In some cases, the epidermides are not in contact throughout their entire surfaces, and whenever this is the case the characteristic epidermal cells reappear, as in *Liliaceæ* and *Amaryllidaceæ*. Similarly, as soon as the carpels of Hellebore become free, the epidermides of the margins appear in their proper character, which now cohere only by contact. It is the same with the axile placentas of the Lily.

As instances where the axis *seems* to be more decidedly prolonged up the centre, are *Lychnis* and allied members of the *Sileneæ*. Ph. Van Tieghem has also shown how an axial cylinder ascends up the middle of the flower of *Campanula medium* for about two-thirds of the height. Thus Fig. 15, *a*, represents a section of the fluted pedicel; *b* shows the lobes isolated, each containing a portion of the fibro-vascular cylinder. In *c*, the broken central cylinder has again closed up, a section showing a complete circle of an axial character. The triangular basal portions of the ovary-cells have now

THE PRINCIPLE OF COHESION. 71

appeared. *d* represents a section of two-thirds of the height of the inferior ovary; but now the fibro-vascular cylinder is dissociated, and forms fifteen separate cords—two being marginal to each placenta and one belonging to each septum. As the cords have their spiral vessels reversed in position, *i.e.* facing outwards and not inwards towards the centre, their axial character has ceased.*

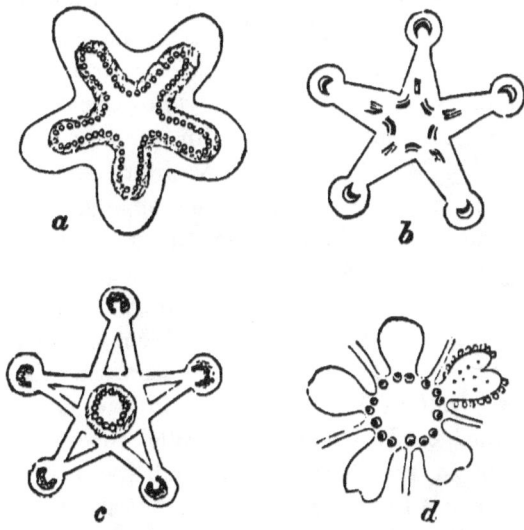

Fig. 15.—*Campanula medium* (after Van Tieghem).

The rule appears to me to be that as soon as, or even before the level of the insertion of the ovules is reached, the internal position of the tracheæ is abandoned. This is the case with *Lychnis*.

In some cases there is an apparently axial formation,

* I do not find that matters can be really expressed quite so "diagrammatically" as, *e.g.*, in his figure *d*; for Van Tieghem does not pay much attention to the *central and scattered positions of the tracheæ*, which I take to be quite as significant as their *outward orientation*; for as the ovules are approached they become dispersed, though a medulla remains.

which has proved to be misleading. Thus, in *Geranium* and allied genera, the beak-like process from which portions of the carpels separate when ripe is not axial at all, but simply the coherent placentas of an entirely carpellary origin.* This will be understood from the description I have given of *Pelargonium* (p. 65).

The mericarps of the fruit of an umbellifer are also supported on a carpophore, which is likewise usually described as axial; but anatomical investigations do not warrant the conclusion. The commissural surfaces are obviously merely the result of rupture between the two carpels which have cohered; and, in consequence of this union, each epidermis fails to develop its true character, but remains in an arrested condition, having the cells somewhat smaller than the rest of the ground tissue. This enables the mericarps to separate readily on maturity. A double fibro-vascular cord runs up the centre to supply ovular cords at the summit of the ovary-cells. If one traces the cords from the pedicel, there will be found in the latter a complete fibro-vascular cylinder. This spreads out at the base of the inferior ovary into ten clearly defined cords which run parallel to each other from base to apex, to furnish the petals and stamens; while two only coalesce and form the axial cord. It is this cord which constitutes the stylopod when the fruit is ripe. Hence it is not axial, but simply the combined marginal cords of the two ovary-cells.

FREE CENTRAL PLACENTAS.—The position of an ovule or ovules on a central support, free from the wall of the ovary, or directly on the base of the chamber, and apparently quite

* Prof. A. Gray (*l.c.*, p. 213) and Henfrey (*El. Course of Bot.*, 4th ed. p. 100) both speak of it as axial; though it was quite correctly described and figured by M. Seringe so long ago as 1838 (*Mem. sur la Fruit des Géraniacées*) : "Les bords de chaque carpel placentaires sont restés et forment la colonne."

THE PRINCIPLE OF COHESION. 73

central, has given rise to a good deal of discussion. Two views have been taken, one being that such ovules are, in some cases at least, axial in their origin, and not carpellary at all; others would refer all ovules, without exception, to a carpellary source. Analogy, indeed, would, if taken alone, seem to justify the latter conclusion, since the numerical proportion of ovules having a decidedly carpellary origin is unmistakably very great; and any doubt upon the matter seems to me to have arisen from a want of due appreciation of the arrest of development, or rather failure of a complete differentiation which has taken place between the ovary and axis at the place where the ovule or ovules appear.

This arrest is particularly apparent, as already stated, in the case of inferior ovaries, as of the Ivy. Thus, in the *Compositæ*, the ovular papilla seems to arise at the base of a cavity in the axis, and might easily be thought to be axial; but a slight eccentricity may be discerned at a certain epoch which is the first indication of its carpellary origin. In *Beta* the basal ovule arises in a very similar manner, but as the ovary becomes more developed, the ovule is carried up so as finally to become pendulous (Fig. 16, *a*, *b*, *c*.) It is much the same in *Typha* and allied genera. The same

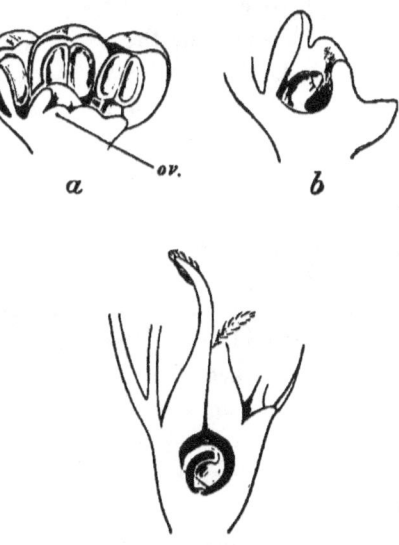

Fig. 16.—*Beta* (after Payer).

gradual elevation of the ovule occurs in *Ricinus* and other Euphorbiaceous plants.

Similarly, if we compare the differences in allied genera, as *Ranunculus* and *Thalictrum;* in the former genus the ovule arises at the very base of the carpel, close to its point of attachment to the axis, and remains there. In *Clematis* and *Thalictrum*, the marginal cleft of the carpel appears a little more decidedly above the base, so that the ovule from its earliest period is situated somewhat higher up, and by a further development is carried to a yet higher position, and so ultimately becomes pendulous. Exactly similar differences occur between the orders *Compositæ* and *Dipsaceæ*.

Hence, it would seem that basilar ovules owe their positions to corresponding degrees of arrest of the growth and development of the carpels, and especially of the basilar portions of the carpellary margins. I think, therefore, we may draw the following conclusion, that the particular form of energy which would cause a carpel to emerge out from and be developed freely and entirely from an axis, is more or less potential than actual.* Consequently, it develops the ovule just where that portion of the carpellary margin *would have appeared* had it been formed; so that the tissue whence the ovular papilla emerges may be considered to be, strictly speaking, neither axial nor carpellary, but undifferentiated merenchyma, and potentially carpellary.

From a single ovule we may now pass to pluri-ovular ovaries. *Dionæa* gives us an instance where many ovules arise at the base perfectly free from the ovarian wall. In this flower the pistil consists of five carpels, which emerge congenitally out of the axis, first as a circular rim, which

---

* It may be noted that it is more *actual* in *Clematis*, etc., in that several ovular papillæ are produced in genera with pendulous ovules, besides being more elevated in position; but only one in *Ranunculus*.

then becomes a cup, which finally contracts above to form the style, just as in *Primulaceæ*. It is, therefore, unilocular, while a circle of ovules appears on a thick ring of tissue within the base of the ovary. Other circles of ovules appear concentrically and centrifugally. It might be questioned, therefore, whether the ring which carries them were axial or not. I think, however, the same interpretation will apply here as elsewhere; that is to say, the ovules arise from the place where the bases of the carpels *would have appeared* had they been differentiated out of the axis.

In the allied genus *Drosera* the placentas are strictly parietal, and the ovules, commencing to emerge half-way up the wall, appear successively, both upwards and downwards. Now, as they are centrifugal in *Dionæa* (corresponding to the *upward* development in *Drosera*), it looks as if only a portion of the upper half of the carpels were really represented at all.

In this genus there is a barren central space within the ring of ovules, perhaps representing the termination of the axis.

That the basal portion only of syncarpous pistils should bear ovules is common enough, and the placentas often swell out there to form bosses which we may reasonably conceive as coalescing to form the continuous ring characteristic of *Dionæa*. Thus *Acer* illustrates how each of the two carpels gives rise to two globular protuberances on which the ovules are borne (Fig. 17). *Anemiopsis*, as figured by Payer, has a confluent protuberance bearing several basifugal ovules. Similar multiovular bosses occur in *Solaneæ* and *Scrophularineæ*, giving the characteristic dumb-bell shape in a transverse section.

Fig. 17.—Carpels of *Acer* (after Payer).

Now, if we imagine these swollen ovuliferous placentas

arising from the basal portions of the carpellary leaves to reach the centre of the ovarian chamber, and be there fused together into a solid mass, we should obtain the apparently axial structure of *Primulaceæ, Santalaceæ*, etc., with the few or numerous ovules *basipetal* in order of development, corresponding to the centrifugal order in *Dionæa* and the ascending order in *Drosera*.

The probability that this is the correct view is supported by a case I have met with in which the carpels of *Primula sinensis* were dissociated, and more or less foliaceous with rudimentary ovules, not only along the margins, but with several borne on heel-like processes,* which extended towards the centre of the ovary, as represented in Fig. 18.

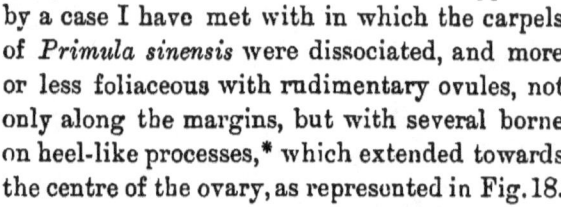

Fig. 18.

Anatomical investigations entirely corroborate the carpellary nature of the central placenta of *Primulaceæ*. The circle of cords, usually ten in number, which pass up the column to nourish the ovules are originally separated from the *sides* of the sepaline by radial chorisis, and become superposed to the sepals; the dorsal cords (about ten) having also parted company from the five sepaline and five petaline. The latter, however, do not give rise to any placentary cords; hence there are really five carpels superposed to the sepals.

With regard to the position of the spiral vessels, they are not oriented as if axial, but are completely embedded in the phloëm, and consequently central. Moreover, the cords in section are circular in form, and not wedge-shaped. The central (if not external) position of the trachæ and the circular form of the cords are both eminently characteristic

* Van Tieghem, though once regarding the central placenta as axial (*Recherches sur la Structure du Pistil*, 1868), has more recently arrived at the same conclusion as myself (*Traité de Bot.*, 1884).

features when they first cease to be axial and become appendicular. The accompanying diagrams (Fig. 19), (a) *Lysimachia nemorum* and (b) *Primula veris*, will illustrate these

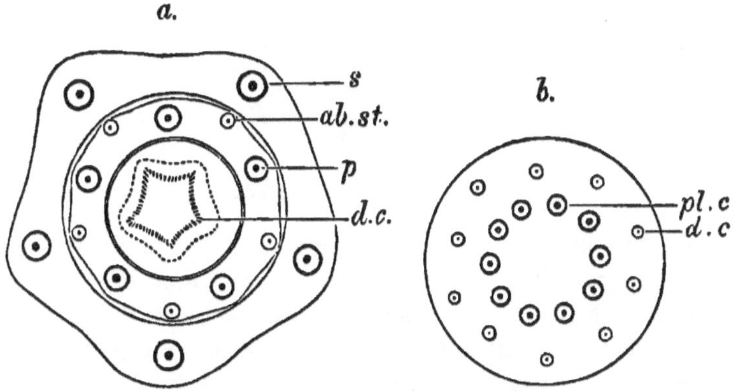

Fig. 19.—a, *Lysimachia nemorum*; b, *Primula veris*.

remarks. The sections are taken on planes * where the pistil is emerging from the receptacle; *s.* represents the sepaline cords; *ab. st.* abortive staminal cords; *p.* the petaline and staminal (combined); *d.c.* dorsal carpellary; *pl. c.* placentary cords.

A free central placenta may result from the destruction of the septa of an originally axile placenta, as occurs in the *Caryophylleæ*. Thus, the ten rows of ovules in *Lychnis* sufficiently indicated their marginal origin. I may add that a careful investigation into the origin and distribution of the cords has convinced me that the axis in flowers of the *Caryophylleæ* early ceases to take any part in the structure of the pistil.

* Fig. *a* represents a section taken rather lower down than in Fig. *b*; as the cords in the latter are still undifferentiated in Fig. *a*.

CHAPTER VIII.

THE PRINCIPLE OF ADHESION.

ADHESION OF ORGANS.—This term is distinguished from cohesion by limiting its application to the union of different whorls. Thus, if the petals or stamens be united to the calyx, they are called episepalous, a term usually syronymous with perigynous; and if the stamens be adherent to the perianth or corolla, they are epiphyllous or epipetalous respectively, sometimes also described as perigynous. On the other hand, if the stamens and pistil be in close conjunction, showing an adhesion between the filament and the style, so that the anther and stigma are brought together, the term gynandrous is applied to them.

Adhesion may be safely regarded as an advance upon cohesion; and there is, I think, a great probability of its being—perhaps, originally, in most if not all cases—a result of adaptation to insect agency.

With regard to the perigynous condition which involves a more or less degree of adhesion of the petals and stamens to the calyx, this is in many clearly a result of the development of the receptacular tube with its honey-disk lining it, as in *Rosaceæ*. This causes the free portions of the petals and stamens to be carried away from the central axis, and placed in a ring "around the pistil," *i.e. perigynous;* while the more or less amount of adhesion of them to the calyx has suggested the term *episepalous*. In the Rose, however, which secretes no honey, the sepals are almost, if not entirely free, and articulate readily; whereas, in other rosaceous plants, if the receptacular tube does not itself fall off, as in *Prunus*, the calyx remains persistent.

Although it is usual to regard perigynous petals and stamens as episepalous as well—that is, "upon the sepals" —when the receptacular tube is well pronounced, it is more strictly in accordance with anatomical structure to regard the former as brought into close proximity to the calyx, rather than being really inserted upon it. In many other cases, as in *Lythrum* and *Daphne*, the whole of the tube has all the appearance of being truly calycine and not receptacular; so that "episepalous" will then best describe their condition of adhesion.

It is rare to find a gamopetalous corolla adhering to the calyx, but it is so in *Cucurbitaceæ*, as in the genera *Cucumis* and *Bryonia*, where the two outer whorls are united.

Ph. Van Tieghem observes * that the union may be the result of the fusion of the respective parenchymas alone, leaving the cords proper to each organ distinct. I think, however, that it will be found to be more frequently the case that when the cords are superposed, they are fused together below, but separate when the organs become free. This is well seen in *Prunus*. The sepaline and petaline cords branch, by tangential chorisis, about half-way up the receptacular tube, and thus give rise to ten stamens. Each of the petaline cords branches on either side again, at a different level, by radial fission, and gives rise to ten more.† So that if we retain the term "episepalous" for the stamens, we must understand that, while the actual stamen is practically free

---

* *Traité Botanique*, p. 390.
† This will be described more fully below (see Fig. 28, p. 95).

from the calyx, yet its cord is common with that of the latter below.

The epiphyllous or epipetalous condition of the stamens is almost invariably associated with a state of cohesion of the perianth-leaves and petals of the corolla; as exceptional instances are *Scilla* and *Lychnis*, which have the parts of the perianth and corolla free, but with the stamens adherent to them; while, conversely, *Campanulaceæ* and *Ericaceæ* have gamopetalous corollas, but the stamens not adherent to them.*

The rationale is primarily, in many, perhaps in every case, an adaptation to insect agency. In the majority of gamopetalous corollas, the honey usually lies somewhere between the insertion of the corolla and pistil, being secreted by one or more glands or an annular disk round the base of the ovary. There are two positions in which the anthers may be placed in regular gamopetalous flowers with reference to the visits of insects for the honey; either around the tube, as in the Primrose and *Scilla*, or close around the style, as in *Convolvulus*, *Campanula*, and *Crocus*. In the former case, when an insect passes its head or proboscis down the tube, it touches the anthers on one side of it and the stigma on the other; but as the proboscis may pass on either side of the pistil in the same and different flowers, that is on the near or remote side, with reference to the position of the insect, such flowers have every facility of being crossed. If they be heterostyled, as the Primrose, then of course each kind has the greater chance of being crossed by the other sort.

* The distribution of the cords in the floral receptacle of *Azalea*, between the insertion of the corolla and pistil, is very anomalous, having no symmetrical arrangement around the centre; while the cords of the corolla of *Campanula*, as described above, are peculiar for other reasons. This may, perhaps, have something to do with the exceptional freedom of the stamens from the corolla.

THE PRINCIPLE OF ADHESION.

In the case of *Crocus*, *Convolvulus*, and other flowers with a contracted base to the corolla or perianth, the anthers are situated close round the style. In these flowers, the insect alights on the stigmas, as already described, grasps the central column and sucks the honey head downwards, and so gets dusted on the abdomen, the pollen from which is thus transferred to the next flower visited.

The adhesion of the stamens to the corolla or perianth thus seems to give a rigidity and firmness, as well as leverage in some cases, so that the action of the insects is more accurately secured, and some one particular spot on their bodies invariably struck and dusted with pollen; which would scarcely be the case if the filaments were free and at liberty to oscillate or swing about in any direction.

In many flowers with irregular corollas, the stamens are declinate; and their adhesion to the tube is then of manifest advantage, for the basal part of the filaments thus acquires an additional strength to act as a fulcrum, which enables the filaments to support the weight of the insect. In *Echium*, for example (Fig. 20, p. 82), the corolla is even strengthened by a rib where the stamen is inserted. This part constitutes the fulcrum. The line of force from the fulcrum intersects a line perpendicular to the filaments, corresponding to the weight of the insect; while the third and upward force is that exerted by the filaments to counteract the resultant of the two former.*

The origin of the adhesion between the stamens and the outer whorls is revealed by anatomical investigations; for the rule is, as described in the case of *Prunus*, that the fibro-vascular cords of the stamens arise by division from those of the outer whorls whenever they are superposed to them.

In other words, when adhesions are seen between the floral whorls, by being superposed to one another, then a

* See also Figs. 38, 39, and 40, pp. 124–126, and consult text.

82   THE STRUCTURE OF FLOWERS.

fusion of their respective cords will be found. If the members arise freely, as in *Ranunculaceæ* and *Cruciferæ*, then their cords are inserted into the axis, having arisen by *radial* division or lateral chorisis.

In the case of the gynandrous pistil, the stamens have their fibro-vascular cords more or less imbedded in the recep-

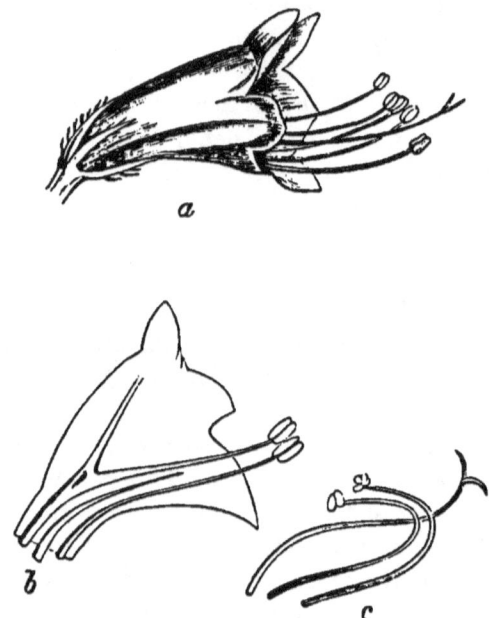

Fig. 20.—Echium; *a*, side view; *b*, before, and *c*, after shedding pollen; showing protandry.

tacular tube, or rather the common tissue resulting from the fusion of the ovary and the tube together; the anther then stands on the summit, and if there be a short or no style, but only the stigmas terminating the ovary, then the anther is in close contact with it, as in *Hippuris*, *Orchis*, etc. When there is a style, the filament may be prolonged in adhesion with it, as in most orchids possessing the so-called column. It is not so, however, in *Aristolochia*, according to Van Tieghem, though often described as such.*

To summarize the above remarks, it seems clear that all adhesions between the two whorls of the perianth, to be found mostly in the *Calyciflorae*, is an accidental occurrence due to the hypertrophied condition of the axis in forming a receptacular tube; so that the term "perigynous" is more strictly applicable than "episepalous."

Adhesions between the filaments and corolla, or calyx if the former be wanting as in *Daphne*, is an adaptation to insect fertilisation; whereby a more rigid position is acquired for the stamens, coupled with a gain of leverage, etc.

Lastly, adhesions between the stamens and pistil only occur where there is a receptacular tube, or "disk," as in *Nymphaea;* and the fusion of filaments with the style, or between anthers and stigmas, is brought about by the very close proximity of the organs when in an early and undifferentiated state.

* Duchartre, *Elém. de Bot.*, p. 648; Henfrey, *l.c.*, p. 125; Benth. and Hooker, *Gen. Pl.*, vol. iii., pt. 1, p. 123; Van Tieghem, *Traité de Bot.*, i., p. 422.

Van Tieghem's description and figure (Fig. 21) is as follows:— "The styles and stigmas are abortive, and the six carpels are reduced to their ovaries. It is, then, the thickened connectives of the anthers, coherent laterally into a tube and covered above with stigmatic papillae, which now play the part of stigmas and of the style."

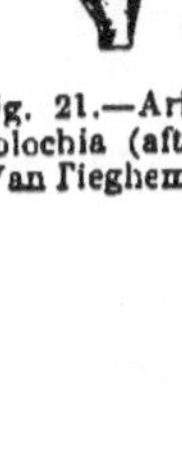

To judge from Payer's figures (*Organogénie*, pl. 91 and pl. 109), the stigmas appear to rise *from the inner side of the very short filaments*, and might be interpreted as truly carpellary stigmas, but fused to the former. A further investigation of the distribution of the fibro-vascular cords should be made. Moreover, *Asarum* does not appear to have anything so abnormal.

Fig. 21.—Aristolochia (after Van Tieghem).

CHAPTER IX.

THE CAUSE OF UNIONS.

Having now noticed the different kinds of unions, we may ask what has brought them about.

We have seen how progressively complex conditions can be traced from entire freedom, as in Buttercups, through forms of Cohesion, such as the gamosepalous, gamopetalous, monadelphous conditions, etc.; to cases of Adhesion, as of the perigynous and epipetalous states; and, lastly, to the adhesion of the ovary to the receptacular tube.

As stated above, these conditions are correlated with greater and progressive differentiations of the floral organs, which have been brought about by insect agencies. The above-mentioned and other terms do not, however, explain how or what the immediate influences are which induce unions of various kinds amongst the parts of flowers; but some researches of Mr. Meehan on the *Coniferæ* * will perhaps give us a clue. There is a well-known and a very generally prevailing feature amongst certain genera of Conifers—as of the *Cupressineæ*, for example—that the foliage can appear under two forms, the leaves being either free from their bases, or more or less adherent to the axis. The two forms of leaves have been recognized as specific characters in *Juniperus*,

---

* *On the Leaves of the Coniferæ*, Proc. of the American Association for the Advancement of Science, 1869, p. 317.

Retinospora, etc.; but both kinds of foliage not infrequently appear together on the same plant; and, when this is the case, the spinescent and free leaves are borne on relatively less vigorous branches, the adherent foliage being characteristic of the more vigorous and quick-growing terminal shoots. It has been also noticed by Dr. M. T. Masters that not only do the broad and free leaves of *Juniperus* and *Retinospora* not occur on the leader shoots, but when the plant is variegated then free leaves (on the stem with arrested growth) are much more variegated than they are on the quick-growing leader shoot.* The last-mentioned observer has also noticed that the free foliage is characteristic of the younger condition of the plant, the adnate foliage that of the adult state.

The conclusions arrived at by Mr. Meehan are as follows: (1) The true leaves of *Coniferæ* are usually adnate with the branches. (2) Adnation is in proportion to vigour in the genus, species, or in the individuals of the same species, or branches of the same individual. (3) Many so-called distinct species of *Coniferæ* are the same, but with their leaves in various states of adnation.

Another very common form of adhesion, to which I have already alluded and which is most probably due to hypertrophy through succulency at an early stage, is fasciation.† Under this condition the fibro-vascular cylinder of at least two "axes," which would be normally separate, coalesce, and form an oval cylinder with, it may be, only a slight

---

* *Gard. Chron.*, 1883, vol. xix., p. 657.

† For remarks on this phenomenon the reader is referred to Dr. Masters's *Teratology*. It is particularly common in herbaceous plants, as Lettuces, Asparagus, etc., and not unfrequent in Ash-trees. I observed a trailing plant of *Cotoneaster* growing over a rockery by the side of a stream in a garden, almost every branch of which was fasciated.

constriction indicating the union. The medullas, cortical and epidermal layers, are also continuous throughout and common to the whole.

Now, the union of two opposite "appendages" to an axis, as in the case of connate leaves, may take place. This may be called foliar fasciation in which the fibro-vascular cords of each "leaf" are embedded in a common parenchyma, and all encased together within a common epidermis.

If we regard the receptacular tube of, say, *Fuchsia* and *Narcissus* in the same light, though adherent to the ovary like a decurrent leaf of a thistle or *Sedum*, I see no argument against the supposition that the tube, in such cases as these, may be regarded as the fasciated petioles of the sepaline and perianthial leaves, now adherent to the ovary within them.

A pear would seem to combine both axis and petioles, as the base of the ovaries is situated much above the commencement of the expansion of the pedicel (see Fig. 22, p. 90, and Fig. 26, p. 94, and consult text).

Each case must, however, be interpreted on its own merits; and I think there will be little difficulty about this, if we recognize the fact that both the pedicel and floral receptacle on the one hand, and the petioles or their floral equivalents on the other, can alike assume all the features of the so-called receptacular tube.

Now let us apply these principles of union through hypertrophy to flowers, and we have an interpretation according to the theory advanced in this book: that differences of floral structure depend largely upon different distributions of nutrition in the several organs; and that the irritation set up by insects themselves is one of the most potent causes of a flow of sap to certain definite places, which encourages local growths, thereby inducing these unions to take place between the parts of any whorl, forming "cohesions," and also between different whorls, or "adhesions."

Other causes may determine them, for hypertrophy may set in through a purely vegetative stimulus; for it is not unfrequent to see abnormal cohesions and adhesions in cultivated orchids, such as petals or sepals adhering to the column, etc. Such may, with a good deal of probability, be referred to the artificially stimulated conditions under which they are grown. These abnormal cohesions between members of the perianth, and adhesions to the column, have been observed both in this country and America.* As a particular instance of the latter kind, Mr. Meehan had observed several dozens of flowers of *Phaius grandiflorus* which had the dorsal sepal united to the column, all being confined to separate spikes from those which have perfect flowers. In some cases, of the same plant two of the petals were united so as to form a hood over the column.

Another peculiarity of Orchids is the tendency to convert sepals or petals into labella, and to multiply the spurs when an orchid is characterized by them so as to render them peloric, a sure sign of hypertrophy.†

All these "monstrosities" seem to point to an excessively unstable condition of equilibrium in the flowers of Orchids; and that they are peculiarly sensitive to the effects of nutritive stimuli, whether brought about by visits of insects or by artificial cultivation. So that the order *Orchideæ* is particularly interesting, as furnishing indirect or even direct

* As by Mr. T. Meehan. *Proc. Acad. Nat. Soc. Phil.*, 1873, pp. 205, 276.

† The remarkable influence of the presence of a "plant-bug," causing the normally irregular corolla of *Clerodendron* to become hypertrophied and peloric, will be described hereafter (p. 130).

proof for my theory—that the forms and structures of flowers are the direct outcome of the responsive power of protoplasm to external stimuli.*

* We may, perhaps, see some analogy between these unions amongst floral organs, which thus occur abnormally in orchids and normally in so many flowers, and inflammatory adhesions in the human subject. It is well known that certain, otherwise abnormal, unions may be congenital, which usually only occur through inflammation set up by abnormal excitation, but they are not hereditary.

I have alluded to hypertrophy and atrophy as causes of the structures of flowers, and shall have more to say about them. I would here add the following analogous phenomena between the animal and vegetable kingdoms. Sir James Paget remarks:—" Constant extra-pressure on a part always appears to produce atrophy and absorption; occasional pressure may, and usually does, produce hypertrophy and thickening. All the thickenings of the cuticle are the consequences of occasional pressure; as the pressure of shoes in occasional walking, of tools occasionally used with the hand, and the like: for it seems a necessary condition for hypertrophy, in most parts, that they should enjoy intervals in which their nutrition may go on actively" (*Lect. on Surg. Path.*, i., p. 89).

The reader will perceive the significance of this passage when recalling the fact that insects' visits are intermittent.

Atrophy by pressure and absorption is seen in the growth of embryos; while the constant pressure of a ligature arrests all growth at the constricted place. On the other hand, it would seem to be the persistent contact which causes a climber to thicken (see p. 156).

CHAPTER X.

THE RECEPTACULAR TUBE.

THE CALYX OR RECEPTACULAR TUBE.—This organ consists of a cellular sheath of varying degrees of thickness, free from or adherent to the ovary. Much discussion has arisen as to the true nature of it, whether it should be regarded as axial or foliar. The older view generally maintained was that it consisted of the lower part of the outermost whorl of the perianth or calyx—in other words, that the basal or petiolar portions of the sepaline leaves were coherent; and if the ovary were inferior, then they were supposed to be adherent to the latter as well.

Schleiden appears to have been the first botanist who propounded the view that it was axial and not foliar. He was followed by others; but this idea took two forms. According to one, it was thought that everything below the summit of the inferior ovary—that is to say, the outer wall, the septa and placentas—was axial, and only the free portion of the summit of the ovary, together with the styles and stigmas, were foliar. According to the other view, it was maintained that the ovaries, styles, and stigmas were foliar, and the superficial covering to the ovary alone was axial. The first view was held by Schleiden, A. de Saint Hilaire, Trécul, Payer, Prantl, and Sachs;* the latter by Decaisne.

* *E.g.* Sachs' *Text-Book of Botany*, Eng. (2nd) ed., p. 566.

Naudin, Ph. Van Tieghem, and, I think, English botanists in general.*

There are three methods of investigation, which conjointly may guide us to the discovery of the real nature of the tube. The first is that of following its development; the second is teratological, and the third anatomical.

MORPHOLOGICAL INVESTIGATIONS.—In tracing the morphological development of flowers of the *Rosaceæ*, where the receptacular tube is a characteristic feature, one notices how a border, surrounding the domelike termination of the axis which soon produces carpellary papillæ, rises upwards and elevates the sepals and the papillæ of the petals and stamens. This border ultimately forms the tube; and the question is, whether it should be regarded as the basal part of the calyx or a development from the axis.

In the *Pomeæ* we find the apocarpous condition of the pistil, characteristic of all the other members of the *Rosaceæ* still retained at first; but in consequence of the growth and close proximity of the tube with the carpels, various degrees of adhesion are brought about between them; thus, in *Pyrus* (Fig. 22, *a*), the bases only of the carpels are from the first fused into the axis. In *Cotoneaster* (*b*) the fusion extends to a higher level on the ovaries. Such "half-inferior" ovaries occur in other genera, as *Saxifraga granulata*, *Gloxinia*, etc. From such we pass to completely inferior states, as in *Compositæ*

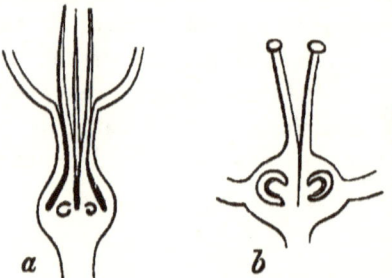

Fig. 22.—*a*, *Pyrus*; *b*, *Cotoneaster* (after Payer).

* Bentham and Hooker describe the inferior ovary of the *Pomeæ* in the terms, "Calycis tubus ovario adnatus."

and *Umbelliferæ*, while *Onagraceæ* furnish illustrations of an extension of the receptacular tube to considerable distances beyond the summit of the ovary, as in *Circæa*, and probably *Fuchsia* and *Œnothera* are similar cases. A like prolongation is seen in some *Compositæ* with "stipitate" pappus, as the Dandelion, *Tragopogon*, *Hypochæris*, etc.

In tracing the development of the inferior ovary of the *Compositæ*, the cavity of the ovary appears to be sunk below the level of the first emergence of the corolla and stamens; and it is this which has suggested the view that the ovary is part of the axis, and that only the style and upper portion of the ovary which is exposed is foliar.

On the other hand, since there are abundant cases of transitional conditions; as, for example, between species of Saxifrage,—*S. umbrosa* having an entirely superior ovary; *S. granulata*, one that is half-superior, and *S. tridactylites*, a completely inferior ovary; and moreover, if we compare the *Pomeæ* with the other tribes of *Rosaceæ*, comparative morphology does not tend to favour the above view held by Sachs, but rather inclines one to the impression that the basal part of the ovary must be carpellary and not axial, though there may be no visible line of demarcation between the cauline and foliar structures.*

The existence of the above-mentioned facts, and many cases of reversion to entire freedom by "solution," supply good reasons for believing that the development of the carpels is *more or less arrested* below, wherever they are in contact with the receptacular tube; yet they retain their power of developing at least one ovule, as is often the case in

* To regard the septa of an inferior ovary "as the prolongations of the margins of the carpels downwards on the inside of the ovary" (Sachs' *Text-Book*, p. 567), seems to be a very strained interpretation in order to fit the axial theory.

gamopetalous epigynous orders. Moreover, the ovule is not strictly basilar and central, but is really situated laterally. Anatomical investigations, as we shall see presently, entirely confirm this view.

TERATOLOGICAL INVESTIGATIONS.—Teratological evidence of the axial, or in some cases, perhaps, petiolar nature of the so-called receptacular tube is tolerably abundant. Thus, in monstrous forms of flowers normally possessing inferior ovaries, the pistil is sometimes completely arrested, when the latter is replaced by a long pedicel which is usually wanting or else is very short, as in Honeysuckle, *Epilobium*, *Orchis*, etc. (Fig. 23).* Pears not unfrequently furnish similar instances, as in the case of the so-called "Bishop's Thumb Pear, which sometimes occurs of an elongated form, destitute of core and seeds. These fruits, which are merely swellings of the flower-stalk, are produced from the second crop of blossoms, which have not energy enough to produce carpels (core) with ovules or ripe seeds."† There is little doubt that the receptacular tube is, in these cases, converted into the rodlike structures in consequence of the total absence of the carpels from within it. In other words, it is axial.

Fig. 23.—*Orchis Morio*, malformed.

There are other indications of the tube being axial in its nature rather than foliar; thus, it frequently becomes "proliferous;" that is to say, flowers, or even branches, may grow out of it, as is often the case with Roses, Prickly Pear, *Umbelliferæ*, etc.‡ Again, certain kinds of Pears, Medlars,

* *a* is the interior of the flower, consisting of a cup-like depression with *two* anthers.
† *Gardener's Chronicle*, Oct. 9, 1886, p. 464.
‡ *Teratology*, p. 100, *seq.*

THE RECEPTACULAR TUBE.

Roses (Fig. 24), etc., occasionally bear foliage on the external surface of the tube, and when the calyx of the Rose becomes abnormally foliaceous, stipules (Fig. 24, *st.*) may appear at the summit of the tube, indicating that

Fig. 24.—Leaf-bearing receptacular tube of Rose (after Masters).

Fig. 25.—Hawthorn with supernumerary free carpels (after Masters).

point to be the base of the sepal. Sometimes supernumerary carpels are borne freely on the top, as in the Hawthorn (Fig. 25).

On the other hand, a tendency to hypertrophy is sometimes discovered in the petioles of leaves of Apples[*] and Pears (Fig. 26, p. 94); and a not infrequent monstrosity is seen in Fuchsias, where one or more of the sepals become foliaceous, and then their petioles are formed but often remain more or less adherent to the ovary if present, which seems to imply that the tube in this plant might be formed

[*] Mr. Meehan describes a similar instance of an Apple-tree which never bore *flowers* but always had an abundance of *fruit*. The latter, however, were composed of metamorphosed and fleshy floral whorls. He adds, however, that cork-cells were formed abundantly on the outside of the apples; remarking, "It would seem, therefore, that with the lack of development in the inner series of whorls necessary to the perfect fruit, those which remained were liable to take on somewhat the character of bark structure" (*Proc. Acad. Nat. Sc. Phil.*, 1873, p. 99).

94　THE STRUCTURE OF FLOWERS.

by, or at least is homologous with, the petiolar portion of the calycine leaves (Fig. 27).

Fig. 26.—Pear with hypertrophied and sub-fasciate petioles.

Fig. 27.—Fuchsia with foliaceous sepals and petals (after Masters).

Phyllomes, however, are after all but modified portions of caulomes, and petioles are still less departures than are blades from the nature of an axis; so that while in some cases one is inclined to regard the tube as more strictly axial, in others it seem to be more homologous with a sort of *fasciation of petioles.*

We shall see directly that the receptacular tube of *Prunus* contains the basal portions of the cords proper to the calyx and corolla, so that we might regard the latter as, on the one hand, axial cords preparatory to forming the perianth; or, on the other, perianthial cords not yet differentiated into petioles.

Similarly, in the case of monocotyledonous flowers, as the Daffodil, since petioles are less differentiated from blades in this class than in Dicotyledons, the inferior ovary may be due to the combination of the pistil with the united sheath-like portion of the perianth, which is prolonged above the summit of the ovary just as it is in Fuchsia, though it is not so prolonged in the Snowdrop.

ANATOMY OF THE RECEPTACULAR TUBE.—Tracing the course of the fibro-vascular cords from the pedicel below the flower, say of *Prunus Lauro-cerasus*, the common laurel, there will be found to be ten, corresponding to the sepals and petals. The cortical tissue and epidermis are continuous throughout, from the pedicel to the summit of the tube. It is well seen also in the tapering end of a pear, from which the cortex gradually widens, while the fibro-vascular cords run vertically up the middle. Before the cords arrive at the border of the free tube of the Laurel, they have given rise to the staminal cords by chorisis, as shown in Fig. 28, *a*, *b*. Fig. *a*

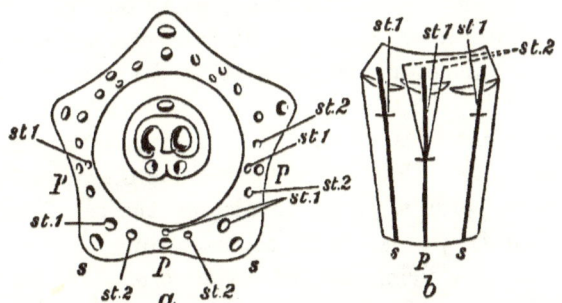

Fig 28.—Receptacular tube of *Prunus* (after Van Tieghem).

represents a section near the edge of the tube in which both the sepaline (*s*) and the petaline (*p*) have given rise by tangential chorisis to a whorl of stamens (*st.* 1); but the petaline by radial chorisis to another whorl (*st.* 2), *i.e.* to twenty stamens in all. Fig. *b* represents a vertical view of the same.*

* The single carpel is represented in Fig. *a* to show the position of its three cords, one being dorsal, and the other two marginal.

As long as the cords are simple, *i.e.* up to the horizontal lines in Fig. *b*, there is nothing to distinguish them from cords of an axis, as in the pedicel. If, therefore, we regard the branches above those levels as belonging to the floral whorls, then the "axis" would terminate at different heights up the receptacular tube—which would seem to be rather too forced a view to be acceptable.

Hence it would seem preferable to regard it entirely as axial until the portions of the perianth issue freely from the upper part of it. We might compare these branches of the fibro-vascular cords embedded in the axis to those belonging to ordinary leaves, which traverse the stem for various distances downwards till they ultimately vanish; only in the case of leaves they are not coherent into a common cord below, but remain free from each other. Moreover, other members of the *Rosaceæ* show that they cannot be always petiolar; because in the rose the sepals reveal their foliaceous character, first by always bearing rudimentary leaflets, and sometimes stipules as well at the top of the tube (Fig. 24, p. 93).

Further complications in the distribution of the cords sometimes arise. Thus, in the tube of the Cherry, I find that the petaline cords assist in furnishing the calyx-limb with vascular cords, just as those corresponding to the arrested stamens of the Primrose enter the corolla of that plant. They either do not branch till they reach the angle between the sepals, or else from a point lower down. The small secondary branches are mainly directed outwards towards the margin, as represented in Fig. 29; *s* being sepaline, and *p* the petaline cords.

In examining transverse sections of inferior ovaries, what one almost invariably observes is an inner epidermis, on some part or parts of which are placentas with ovules,

THE RECEPTACULAR TUBE.

an outer epidermis, and an intermediate ground tissue, apparently nearly uniform in character, from one epidermis to the other (as in Fig. 14, *a* to *e*, p. 68). A definite number of fibro-vascular cords penetrates this ground tissue. Theoretically, if this structure consist of two parts, viz. the interior carpels and the exterior "tube," some line of demarcation might be expected to be traceable; but in the majority of cases it would seem that, as neither the *inner* epidermis of the tube nor the *outer* one of the carpels are required, they are not developed at all; and so the internal tissues of the two organs become confluent and uniform, and this accounts for the fact that the dorsal cords at least are simply embedded in this common tissue. Nevertheless, in some cases there actually is a certain differentiation in the tissue, as Van Tieghem has shown in the case of *Alstrœmeria versicolor* (Fig. 30), where a yellow band of cells marks the

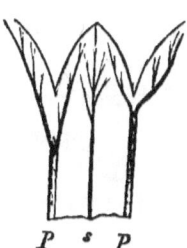 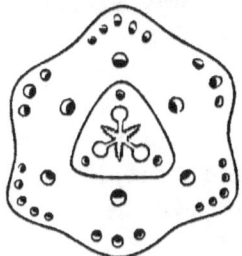

Fig. 29.—Receptacular tube and calyx-limb of Cherry.

Fig. 30.—*Alstrœmeria* (after Van Tieghem).

junction or congenital fusion of the two parts (indicated by the line in the figure).

From the preceding descriptions, it will be seen, with regard to the sources of the cords belonging to the inner whorls, that they arise by division, radial or tangential as the case may be; and then the secondary cords thus parted off are generally included within the tissue of the tube.

These cords of the inner whorls may be given off at the terminal point of the pedicel; that is, at the base of the flower. In this case they may all run parallel from the base to the summit of the receptacular tube; or they may branch at various heights within the tube itself, as in *Prunus*, described above; or, lastly, they may not arise until the summit of the ovary is reached, when they pass off and enter their respective floral organs directly. These variations occur in both free receptacular tubes as well as when coherent to ovaries.

As an example of the first case may be mentioned *Alstrœmeria versicolor;* of the second, *Galanthus nivalis*, or Snowdrop; and of the third, *Narcissus*. In *Alstrœmeria*, all the floral appendages have their cords distinct and independent, but invaginated by the tube of parenchyma throughout (Fig. 30). In the Snowdrop, the carpellary cords are distinct, but the perianth and andrœcium are inserted in the pedicel by a single verticil of cords, which becomes double higher up. Lastly, in *Narcissus*, all the parts of the flower are originally inserted in the pedicel by six cords, of which three give rise by successive tangential fission to a radial series composed of the dorsal cords of the carpels, the stamens opposite to the sepals, and the sepals themselves. Similarly, the other three form the petals together with the whorl of stamens opposite to them.*

In *Campanula*, and to some extent in *Lobelia*, the cords

* Ph. Van Tieghem, to whose researches I am indebted for the above, but which I have also paralleled in other cases, represents them neatly by the following formulas, wherein ( ) signifies vascular union, and [ ] the cellular union of the receptacular tube; while (d) stands for the dorsal and (m) the marginal cords of the carpels. St_p signifies petaline and $St.$ sepaline stamens.

$$Alstrœmeria - [3\ S + 3\ P + 3\ St_s + 3\ St_p + 3\ C_s].$$
$$Galanthus - [3\ (S + St_s) + 3\ (P + St_p) + 3\ C_s].$$
$$Narcissus - [3\ (S + St_s + d\ C_s) + 3\ (P + St_p) + 3\ C_m].$$

belonging to the petals are given off by radial chorisis from the sepaline, either quite from the base of the ovary or from about midway up the tube; they then diverge right or left at an acute angle, and, as soon as they have reached the summit of the ovary, pass up into the corolla.* As a rule, however, the petaline cords of flowers are quite distinct from the sepaline; the six or ten, common to Monocotyledons and Dicotyledons respectively, forming the fibro-vascular cylinder in the pedicel.

In all these and other cases the cords running up the receptacular tube proceed originally from the petiole, and are, so to say, even there *intended* for the appendages above. Normally they retain their axial character, in being arranged in a circle round the centre; abnormally an appendicular character can be revealed, by their becoming free and assuming a foliaceous aspect, as in Roses or *Fuchsia*, as mentioned above; so that as long as the tube is normal, *i.e.* a cylinder of cortical parenchyma with cords, it is of the nature of axis, and can develop extra phyllomes and even buds; but abnormally, the foliar nature, usually limited to the floral members at the summit, is extended to a greater distance lower down and the cords may now be converted into petioles, etc.

Hence it appears undesirable to call it either a calyx tube or axial; for these terms would seem to bind one to consider it permanently and in all cases as being either of one nature or the other. The term receptacular tube is therefore best, as it certainly "receives" or supports the whorls of the flowers; and Teratology clearly shows that it can be either foliar (petiolar) or axial according to circumstances.

* This reminds one of the way in which stipular appendages of *Galium*, etc., are supplied with cords—not by their intercalation into the common fibro-vascular cylinder of the stem, but—from a horizontal circular zone of fibres which connects the cords of the opposite leaves.

Just as the two complete vascular cylinders of two separate floral peduncles can become fused into one oval cylinder when the latter are " fasciated," so, too, would it seem that the cords belonging to the separate parts of a floral whorl, where there is no receptacular tube, can form a single united cylinder, which one then designates as the receptacular tube.

In the case of the inferior ovary, I would again emphasize the fact that the difficulty felt as to what is axial and what carpellary is entirely removed if the undifferentiated condition of the carpels be thoroughly understood. Indeed, whenever two organs are congenitally in union the epidermis of each is undeveloped, and the two mesophyls become one; so that the dorsal cords of the carpels and those proper to the axis are alike plunged into a common tissue, which, regarded as one, is neither wholly axial nor wholly carpellary.

CHAPTER XI.

THE FORMS OF FLORAL ORGANS.

THE FORM OF THE PERIANTH—GENERAL OBSERVATIONS.—It requires but a most cursory observation of flowers to notice how great is the variability in the forms of all their organs; and the questions now before us are, how these morphological characters are correlated to the one process of pollination in order to secure the fertilisation of the flower, and how this infinite diversity of form has arisen.

Most important differences in this respect follow from the fact of flowers being regular or irregular, and, when adapted to insects, according as the honey is easily accessible or not. Regular * flowers when borne singly are almost always terminal;† and when they are arranged in racemes, etc., they either stand out erect at the ends of their pedicels so as to be readily approached at any point of their circumference, as in the Wallflower, or else they are pendulous; under which conditions, as a rule, no particular part is favoured by the

* It is usual to speak of a *flower* as being regular or irregular; but the term should be, strictly speaking, confined to one whorl at a time; though when the corolla is irregular, the calyx and stamens are usually somewhat irregular as well.

† The central and terminal flowers of many plants which elsewhere bear irregular flowers are often regular, as in Horse-chestnut, *Pelargonium*, several of the *Scrophularineæ*, as Snapdragon, *Linaria*, *Pentstemon*, etc.

THE STRUCTURE OF FLOWERS.

insect more than another. It is only when the flower is situated laterally and projects horizontally, or approximately so, with its limb or border in a vertical plane, and, moreover, is more or less closely applied to the axis, that an insect is compelled to alight upon it on one side only, when approaching it directly from the front. It then throws all its weight upon the organs on the lower or anterior side of the flower, as is the case with the keel petals of papilionaceous flowers, with the lips of Labiates, etc.; or else its weight is sustained by the stamens or style, or by both together, as in *Epilobium angustifolium, Circæa, Veronica,* Larkspur, and Monkshood; and whenever the stamens are declinate, as in Horse-chestnut, *Dictamnus, Echium, Amaryllis,* etc.

Flowers which have irregular corollas mostly show various degrees of "bilateral" form in their different whorls, and, have been called "zygomorphic." Such flowers, as a rule, do not receive the visits from so many different species of insects as regular flowers. These latter, not being characterized by the possession of any very definite contrivances for securing special insect agency, are accordingly visited by a much greater number and variety than those flowers which have become markedly adapted, and consequently restricted to particular visitors.

It must not be forgotten, however, that regular flowers, if the tube leading to the honey be very contracted and more or less elongated, may become almost as much exclusive as very irregular ones; for such flowers are mainly restricted to Lepidoptera.

The following examples may suffice to illustrate these facts. *Ranunculus acris,* which is perfectly regular and with no specialized structure, is visited, according to Müller, by more than sixty different species of insects; whereas species of *Aconitum* and *Delphinium,* the two most highly differentiated and the only genera with irregular flowers of the same order, are adapted to, and mainly visited by the larger species of bees. Similarly of conspicuous and regular flowers of *Rosaceæ*, *Prunus communis* has twenty-seven visitors; *Spiræa Ulmaria*, twenty-two; *Rubus fruticosus*, sixty-seven; *Fragaria vesca*, twenty-five; *Cratægus oxyacantha*, fifty-seven. On the other hand, of irregular flowers, *Digitalis purpurea* has only three useful visitors; *Linaria*, nine or more species of bees, and *Orchis mascula* only eight.

As an instance of a long-tubed regular flower, *Lonicera cœrulea* may be mentioned. It is adapted to humble-bees, by which it is chiefly visited. Similarly, the flower of the Honeysuckle, the lobes of which are scarcely if at all unequal, admits only a few lepidopterous insects which can reach the honey. So, too, *Asperula taurina*, which has a tube 9 to 11 mm. long, is visited by nocturnal Lepidoptera.

THE ORIGIN OF IRREGULARITY.—With reference to the theoretical origin of irregular whorls, I assume that they have all descended from regular ones through external influences.* With regard to terminal, regular flowers the flow of sap is directed equally, radially, and in all directions on reaching the floral receptacle, and there is no inherent cause to make a terminal flower zygomorphic, or to induce one or more parts of any whorl to grow differently from the rest. Hence the primary cause of irregularity must come from without, and I regard this cause as issuing from the insect itself; namely, the mechanical influence of its weight and pressures. To this external irritation the protoplasm of the cells responds, and gives rise to tissues which are thrown out to withstand the strains due to the extraneous pressures

* The fibro-vascular cords of the pedicel are arranged at regular intervals, and are perfectly symmetrical around the medulla in irregular flowers, just as they are in the case of regular ones.

of the insect, and so the flower prepares itself to maintain an equilibrium under the tensions imposed upon it, and irregularities are the result. Such, for example, occur in bilobed calyces, as of Furze and *Salvia;* in the many forms of "lips," or labella,* and enlarged anterior petals; in dependent stamens, as of Aconite and *Epilobium angustifolium,* or in the more usually declinate condition, as of *Dictamnus, Amaryllis,* etc. In these latter instances, in which the andrœcium bears the burden, the anterior petal is either, as a rule, unaffected, and shows no increase in size, or else there is a tendency to atrophy, so that it is reduced in size, as are the keel petals in *Amherstia.* It is sometimes even wanting altogether, as in the Horse-chestnut.†

* If the flower be resupinate, then it is the posterior organ which, now being in the front, has become enlarged; as in *Viola* and *Orchis.*

† There has been more than one investigation into the causes of zygomorphism (as by Vöchtung, *Ber. Deutsch. Bot. Gessell.,* iii. (1885), p. 341; and Pringsheim's *Jahrb. f. Wiss. Bot.,* xvii. (1886), p. 297 : also, by Dr. F. Noll, *Arbeit. Bot. Inst. Würzburg,* iii. (1887), p. 315). H. Vöchtung distinguishes three different sets of causes as producing zygomorphism, viz. gravitation only; gravitation acting on the constitution of the organs; and the constitution of the organs alone.

An objection to gravitation pure and simple is, that *all* flowers would be more or less subject to it, and become more or less zygomorphic accordingly. It does not account for the infinite diversity in the forms of zygomorphic organs; nor for the many correlations for insect fertilisation which exist between all parts of the flower. If to gravitation, however, we add the weight of the insect, which simply intensifies it, and couple with this the pressures exerted by the insect in various directions, *then* we have an adequate theory, which gravitation alone could not supply. When Vöchtung speaks of "constitution alone" as a cause, I presume he means *hereditary effect.* If so, I would quite agree with him, as zygomorphic flowers *now* grow to be such from purely hereditary influences. When, however, he would attribute the form of *Epilobium angustifolium* to geotropism, as the supposed cause of the lowermost petals bending upwards, and the stamens and style downwards (see Fig. 34, p. 111), I do not see how

Compensating processes thus come into play, so that while some parts are enlarged others are diminished, the former always having to bear the strains, while the latter are free from them. Thus the lip of *Lamium* consists of one much-enlarged petal, which forms an excellent landing-place, but the two lateral petals, not being required, are atrophied to mere points. Similarly, while the two posterior petals enlarge to form the hood, presumably due to the backward thrust of the insect's head, the posterior stamen has vanished altogether. The gamosepalous calyx now furnishes its aid to support the slender tube of the corolla, not only by doubling its number of ribs, but by uniting them all together by means of a sclerenchymatous cylinder within the mesophyl.

If the tube of the corolla be very strong and well able alone to support the insect, the adhesion of the filaments being also a powerful addition to its strength, then the calyx often remains polysepalous, as occurs in the Foxglove, Snapdragon, *Petunia*, etc.

If, instead of the anterior petal forming the landing-place, the tube of a gamopetalous corolla has enlarged so as to admit the ingress of an insect which partly or entirely crawls into it; then it is this tubular part which, more especially having to bear the strain upon it, bulges outwards, or becomes more or less inflated in form; while the lip or anterior petal, not having to bear the entire burden, is not particularly enlarged, if it be at all. The Foxglove and *Gloxinia*, as well as *Petunia* to a slight extent, illustrate this adaptation in irregular flowers, while "campanulate" flowers afford examples amongst regular ones.

gravitation can act in any other way than "downwards." But if one observes how a humble-bee suspends itself on the stamens while its body, so to say, thrusts the petals aside and upwards, we find a much more satisfactory interpretation in the theory I have proposed.

If no more than the head of an insect enter the flower, then the corolla shapes itself to fit it. Thus Snowberry, *Scrophularia*, and *Epipactis* only admit the heads of wasps, which are the regular visitors of these plants.

Other instances in which the limb is not much, if at all, enlarged occur in flowers especially adapted to Lepidoptera. Hovering, as they generally do, before the flowers, and inserting their long proboscides while on the wing, there is no tendency to develop larger anterior petals, but the irritation affects the tube only, which thus elongates and contracts, resulting in little or no irregularity in the flowers, as in *Œnothera biennis*, in which the calyx tube has contracted, or in Honeysuckle, which has a tubular corolla. If bees or other insects visit the flower as well, then some degree of obliquity may result, as in *Teucrium Scorodonia*.

Thus, then, may we get a rationale of the structure and form of floral organs, and their great diversity corresponds to a similar diversity in the insect world; for the flower, if it be visited by many, will presumably take a form corresponding to the resultant of the forces brought to bear upon it; if visited by few, it will shape itself in accordance with the requirements of its principal visitors; and thus is it that while some easily accessible flowers receive many classes of insects, others are restricted to few, or even one; and then the insect and the flower are so closely correlated as to almost impress upon one the idea that they were mutually created for each other!

The accompanying figures of *Duvernoia adhatodoides*[*] may illustrate my meaning. Looking at Fig. 31, *a*, alone (supposing we know nothing of insect visitors), one might ask, For what use is this great irregularity? why and how has it

[*] From a paper by Mrs. Barber, *Journ. Lin. Soc. Bot.*, vol. xi., p. 469.

THE FORMS OF FLORAL ORGANS. 107

come into existence? And no answer is forthcoming. Now turning to Fig. 31, *b*, we see one use at least. The weight of the bee must be very great; and the curious shape of the lip, with its lateral ridges, is evidently not only an excellent landing-place, but is so constructed as to bear that weight. Moreover, the two walls slope off, and are gripped by the legs of the bee, so that it evidently can secure an excellent purchase and can thus rifle the flower of its treasures at its ease.

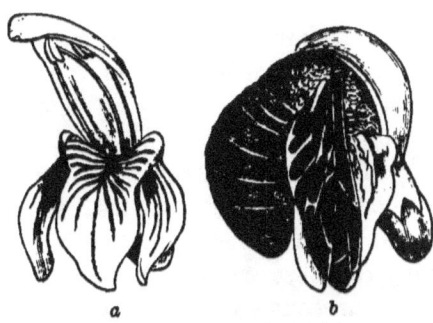

Fig. 31.—*Duvernoia adhatodoides*.

Irregular corollas are very numerous, but certain principles, traceable to insect action, govern their forms. In the first place, the side upon which the insect rests, or at least upon which its weight is thrown, is always enlarged, and mostly forms the landing-place. It is *almost* always the anterior petal; if, however, the pedicel or ovary has been too slender to support it, then it has sometimes become twisted, and the flower is said to be resupinate, so that the posterior petal becomes anterior in position, and is now the larger one, since it supplies the landing-place for insects, as in *Orchis*. *Fumaria* might be called semi-resupinate, as the corolla has only rotated through 90°. A slight modification occurs in the "Bee-orchis," *Ophrys apifera*, which is usually described as having a twisted ovary like a true *Orchis*; but in this species it has scarcely if any twist at all; the flower, however, is bent over to the opposite side of the stem, so that while the posterior petal is still the labellum, the ovary has itself remained perfectly straight.

The next point to notice is that when the anterior petal is enlarged, the posterior one or more often enlarges also, while a corresponding tendency to atrophy affects the lateral ones. This is seen in many species of *Leguminosæ*, *Scrophularineæ*, and *Labiatæ*, and in zygomorphic flowers generally. It occurs thus in the wing petals of many papilionaceous flowers, as is particularly well seen in *Onobrychis*. The immediate causes, I repeat, I would recognize in the weight of the insect in front, the local irritations behind, due to the thrust of the insect's head and probing for nectar, coupled with the absence of all strains upon the sides. In some papilionaceous flowers the wing petals form a landing-place, as in *Indigofera* and *Phaseolus*. Whenever this is the case, they too are enlarged, as the lateral ones are in Fig. 31, and undertake the duty impressed upon them.

When, therefore, one finds as an invariable rule how the front petals enlarge when flowers are compacted and visited only from the front, and thus become irregular; and as such often occur in orders where flowers are normally regular, as *Iberis*, *Centaurea*, *Heracleum*, etc.; and, moreover, when the same phenomena appear in orders having no affinity between them, as in *Labiatæ* and *Orchideæ;* and are, indeed, to be found throughout the length and breadth of the floral world, one is justified in attributing such irregularities to a common cause, that being, according to my theory, the responsive power of protoplasm to the irritations from without, set up by insect and other agencies.

Many other special cases might be described from the different orders of plants, but the above will suffice to illustrate this principle of responsive action with resulting correlations to insect agency. I would here, however, call the reader's attention to the mechanical arrangement of forces as shown in *Lamium* and *Echium*, where it will be seen that the

THE FORMS OF FLORAL ORGANS. 109

adhesions of the stamens to the corolla furnish the fulcra, the cohesion of the petals into a tube affording a greatly increased power of resistance; the weight of the insect on the labellum or declinate stamens is, of course, vertically downwards, and the line of the resultant, which the lip in *Lamium* and the stamens whenever declinate have to exert, passes through the point of meeting of the first two, and so sustains the insect while visiting the flower. Other and analogous instances will be described hereafter.

Good illustrations of the occurrence of great thickenings just where the strain will be most felt, may be seen in the slipper-shaped flowers of *Calceolaria* (Fig. 32), *Coryanthes*, and *Cypripedium*. Thus *Calceolaria Pavonii* possesses a thick ridge along the upper edges of the curved basal part, which carries the inflated end upon which the bee stands, and which it depresses to get the honey. In this species it may be noticed the anther-cells are separated (*a*), so that they can oscillate as they do in *Salvia*. In *Cypripedium* the edge is folded inwards, thus strengthening the same part; while in *Coryanthes* the lower portion is enormously enlarged, thus acting as a powerful spring which forces the anterior end of the labellum to be in close contact with the column.

Fig. 32.—*Calceolaria Pavonii* (after Kerner).

THE ORIGIN OF IRREGULARITY IN THE ANDRŒCIUM.—As it is with the perianth, so is it with the androecium: if the petals are regular the stamens are usually regular also; but when irregularity occurs in the corolla the staminal whorl follows suit, and the position and form of the stamens are equally correlated to the effectual pollination of the flower. Thus, as hypertrophy affects the anterior side of the flowers of *Labiatæ*, the anterior stamens are almost invariably the larger pair. On the other hand, atrophy has affected the posterior side of the staminal whorl, causing the total loss of the fifth stamen, and, to some extent, a reduction in length of the next pair of filaments.

When the weight of the insect is thrown upon the stamens, they either hang downwards, and the insect is suspended upon them, as in *Epilobium angustifolium*, or else they become declinate and then the anterior petal, being relieved, does not enlarge, either remaining of the same size as the rest, or else diminishes, and may even vanish altogether. Thus *Vallota*, with its perfectly regular perianth and spreading stamens, may be compared with *Amaryllis*, which has declinate stamens and a small anterior petal. The terminal flower of a "thyrse" of the Horse-chestnut, like the terminal flower of a "truss" of Pelargonium, is often regular with spreading stamens, whereas the normal flowers have declinate stamens, and usually only four petals, the fifth or anterior one being altogether suppressed.

Fig. 33.—*Dictamnus* (after Tieghem).

In some flowers the stamens are dependent at first, but their anthers rise up when dehiscing, and so the filaments become declinate in the pollinating stage. This is the case with *Delphinium*, *Epilobium angustifolium*, and *Dictamnus* (Fig. 33). In this flower the anterior petal is of much the same size as the others, but is often displaced (Fig. 33), and not immediately below the stamens,—this

THE FORMS OF FLORAL ORGANS. 111

lateral displacement of the anterior petal being not always carried out, as it is in the next flower to be described.

In *Epilobium angustifolium* (Fig. 34) and *Godetia*, which have no anterior petals, the bees cling to the dependent stamens, while the petals have become permanently displaced, the two lower being somewhat raised, so that the angular distances are not the same. In *Azalea* and *Rhododendron* there is no anterior petal, but the posterior one is slightly enlarged, and this alone possesses extra colouring and the "path-finder." The stamens, being declinate, carry the insect without the aid of the corolla, so that the antero-lateral

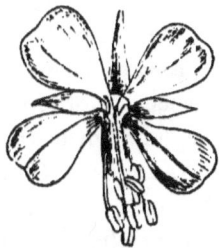 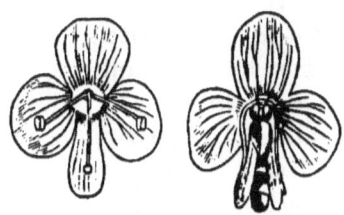

Fig. 34.— *Epilobium angustifolium.*   Fig. 35.— *Veronica Chamædrys* (after Müller).

pair of petals, not sharing in the support of the insect, are not enlarged at all.

In *Circæa* and *Veronica Chamædrys* (Fig. 35), the insect clings to the two stamens and style; and the anterior petals are not enlarged, while in the latter flower it is, as usually the case, the smallest, the stamens of *Veronica* being attached to the lateral petals have to supply the fulcra for leverage, and consequently these have now become relatively hypertrophied.

In many flowers which have sub-declinate stamens, the latter lie in a more or less boat-shaped anterior petal, showing that the action of the insect has somewhat affected both the whorls together, as they have each some share in carrying the insect. Such is the case in the *Ocimoideæ* of *Labiatæ*, in *Collinsia bicolor*, the "Lemon-scented" *Pelargonium*, etc.

CORRELATION OF GROWTH.—I have only referred to the forms of flowers as grouped under the terms "regular" or "irregular," and alluded to a few instances; for it is not my object in this work to merely give illustrations of various kinds, which are presumably well known to the reader, but to offer a rationale of the whole, without, however, attempting to say how each individual shape has actually come into existence. To do this, it would be impossible in the present state of our knowledge of the history of flowers; my object being to suggest a probable cause, namely, the mechanical influence of insects, without excluding others which we cannot trace. Nutrition, however, must be always borne in mind as an important one, hereditary influences as others—as, for example, in the restoration of an irregular flower to a condition of regularity, as occurs in *Linaria*, *Lamium*, *Gloxinia*, etc. The point, however, which I would specially emphasize is the correlation existing between the several parts of the organs, so that, regarded collectively, they all conspire to secure one and the same end, that being the pollination of the flower. Thus, as I have shown above, the calyx of *Salvia* has a form and structure correlated to the tube of the corolla; the corolla has a form in strict adaptation to the weight and pressures of the insect which rests upon the lip. The stamens are, again, correlated to the pressures brought to bear upon them, and have grown in response, forming the remarkable lever-processes, which are also found in species of *Calceolaria*. Lastly, the style and stigma are correlated to the position of the anthers. Hypertrophy in one direction has brought about atrophy in another, so that the two posterior stamens, are rudimentary, while the fifth has vanished altogether.

THE FORMS OF FLORAL ORGANS. 113

Now, it might be argued, that when one organ changes its form others *must* do so in obedience to the "laws of correlation of growth," as Mr. Darwin showed to be the case with the feet and bills of pigeons. In plants, however, the connection between various parts, even in close proximity, is by no means so intimate as between different organs of the higher animals; while the theory advanced here gives a common interpretation for the whole of the so-called correlations found in any flower. That one is justified in saying that correlated growths are much restricted in plants, is clear from the experience of horticulturists; thus, while, *e.g.*, the varieties of pease are infinite, they having been the object of selection alone, the flowers which produce them have virtually remained unchanged.

A single coincidence has little or no scientific weight as indicating cause and effect. It is only when coincidences can be multiplied that they furnish a probability of a high order; which, even if they do not admit of a verifiable experiment, still furnish a *moral conviction*, which, by the rules of philosophy, is equivalent to a demonstration. Now, this is exactly the case with irregular flowers. They always occur in similar positions; they are always constructed so that the insect in adaptation to them can gain access to the honey in the easiest way; their organs are so situated that the pollen should be transferred accurately to the stigma, etc. And when we find them distributed everywhere throughout phanerogamous plants, the probability that the same or analogous causes have brought them about is of a very high order indeed

Moreover, since we have abundant evidence of the responsive power of protoplasm to build up tissues wherever they are required, I am not assuming an influence on the one hand without ample evidence of the probability of the responsive action on the other, coupled, of course, with hereditary and other influences which fix the variation. Thus, then, as I believe, all flowers as we have them now, which are in perfect adaptation to insect agency, are the outcome of the resultant of all the forces, external and internal, which the insect has actually brought into play or stimulated into action by visiting them for their honey or pollen.

The belief that such processes may have grown in response to mechanical irritations is supported by some interesting experiments made by Mr. O'Brien, of Harrow, who has kindly favoured me with the following remarks: " With reference to impressions conveyed by 'nervous' force in Orchid flowers, whereby the expansion of the sepals and petals signifies to the reproductive organs that the time for fertilisation has arrived, I have observed that the periods of maturing and of decay may be either arrested or hastened in certain orchids by artificial means. With reference to arresting decay, I took such flowers as *Stanhopea* and *Coryanthes*, which have large membranous sepals, and which, in the ordinary course of events, become reflexed soon after the opening of the flowers, and shortly afterwards wither. These are then followed by the other parts. By seizing the opportunity as soon as they expand, and by passing a thread round them, so as to keep them in the condition of the flower when just on the point of expansion, they may be kept good for a long time, the flowers evidently, as it were, not realizing the increased lapse of time, and being unaware that they had passed the period when they would have been ready for fertilisation. When so secured, a flower of *Coryanthes speciosa* on my table kept fresh three times as long as it would have done on the plant. The dripping of the water from the horns above the bucket is also arrested. Finally, on releasing the ligature, the broad wing-like sepals immediately became reflexed, and the water commenced to drip. Shortly afterwards the wings shrivelled up, and the flower decayed in the same manner as it would have done a week before if left to itself on the plant.

"I will now give an example of deceiving a flower by artificial means, by making it believe that its fertilisation has been accomplished without its having taken place at all. *Miltonia Russelliana* carefully guards the approach to the column by closing the petals over it; but on pushing these petals aside with a pencil, I always found that the labellum faded, and withdrew upwards very soon afterwards. The showy portion of the flower, evidently having had it conveyed to it that its duty was performed, then followed suit. On carrying the deception still further to the reproductive organs, by placing small pieces of grit on the stigma, I found that the ovaries would swell in many cases, just as though the flower had been properly fertilised by pollen. This same result often takes place in Orchid flowers under cultivation, and seed-vessels are obtained of full size, but, of course, with no vitality in the grains within."

As an analogous instance, I will add that it is the belief of M. O. Beccari that ants are not only responsible for the remarkable growths in *Myrmecodia* and *Hydnophytum*, etc., but that they have become indispensable for the healthy development of such plants. The investigations of M. Treub on *Dischidia*, the pitchers of which are frequented by ants, like the stipules of *Acacia sphærocephala*, seem to justify one in concluding that genus also to be one of these so-called "Ant-plants" (*Ann. du Jard. Bot. de Buitenzorg*, iii., p. 13).

Dr. Lundström also believes that the habit of producing "domatia" is now hereditary, without the actual presence of the insects (see *Journ. Roy. Micr. Soc.* 1888, p. 87.)

CHAPTER XII.

THE ORIGIN OF "ZYGOMORPHISM."

BILATERAL SYMMETRY.—A feature abundantly illustrated through the flowering world, in the construction of irregular flowers which are highly specialized for insect agency, and of which the *Labiatæ* and *Scrophularineæ*, for example, furnish many instances, is the hypertrophy of the corolla in the direction of an antero-posterior plane, giving rise to a bilateral structure.

On the one hand, the lips of various kinds, as also the keel, and often the wing petals too, where they help to support the insects in papilionaceous flowers, are accounted for by the weight of the insects bringing about a responsive action in the protoplasm, thus determining a flow of nutriment to the parts demanding it, which now grow into the forms required. On the other hand, the opposite or posterior side is often influenced as well, so that, as in *Lamium*, the lobes of the two posterior petals have grown into the enlarged hood. The cause of this I take to be the powerful *thrust* which insects exert against the posterior side while their *weight* is expended on the anterior. If a humble-bee be watched, as represented in Fig. 31 (p. 107), it will be seen how eagerly and determinedly it forces its way into a corolla-tube if it expand upwards, as in *Duvernoia* or *Lamium*. All the pressure is exerted along the median plane, like an oblong wedge

THE ORIGIN OF "ZYGOMORPHISM." 117

thrust into a circular tube. The corolla then "gives," as it were, and expands along the antero-posterior plane. The calyx follows suit, and often assumes a bilobed funnel-shaped tube as well; while the lateral lobes of the corolla tend to atrophy, since they do not lie along the line of the pressure due to the weight of the insect (see Fig. 40b, p. 126.)

If the floral organs be imagined to consist of some plastic, extensible, but not elastic substance, and be subjected to various pressures, strains, thrusts, etc., in imitation of the motions of insects, it is readily conceivable how the parts would yield, stretch, or bulge, and become fixed into shapes very closely resembling what has actually taken place in nature. In reality, of course, the ability *to grow in response to the forces applied* is to be substituted for the theoretical plasticity and extensibility of the imaginary material.

Compensatory degenerations occur in various directions, as in the atrophy of the lateral petal-lobes of *Lamium*, the loss of the fifth posterior stamen, the reduction in length of the filaments of the posterior pair of stamens. In this latter respect *Nepeta* differs from other genera, but as we can readily conceive how all sorts of differences may and do exist in the direction and degree of the forces applied to flowers, some exceptional ones must have occurred in that genus which has favoured the growth of the posterior pair, so that they have become the longer ones; for there is no rule without an exception. As another illustration, *Teucrium* may be taken. In this genus the "hood" is entirely wanting; but here, again, the interpretation is that, no hypertrophy having been applied to them, the two petals of which it is composed have become reduced in size and "cleft," as shown in Fig. 36, of *T.* (*Teucris*) *orientale*. Bees,

Fig. 36.—Flower of *Teucrium* (after *Bot. Mag.*, 1279).

when visiting the flowers, hang downwards upon the corolla, as the lip and adjoining lobes are in one vertical plane, and give no thrust upon the posterior side. All weight, therefore, is thrown upon the front, just as it is on the stamens of *Epilobium angustifolium*, described above. Their weight has consequently, so to say, "split" the hood in twain, and the stamens now stand erect in the cleft.

The peculiar form of the corolla, with the whole of the limb dependent in a vertical direction, must throw the weight of the insect so much to the front, that the leverage will be at a considerable disadvantage—much more so than when the insect stands more directly over the tube of a corolla; which latter, in that case, is often strengthened by that of the calyx. To meet this difficulty the pedicel is curved over at the top, as may be readily seen in our common Wood-sage, and forms a spring, while hypertrophy has attacked the posterior side of the calyx, in that it now carries two extra marginal ribs, one on either side of the posterior dorsal one, as shown in the accompanying diagram. This is exactly the reverse of what occurs in *Salvia*, and others which are much more strengthened on the anterior side, when the insect stands more directly over the centre of the flower.

```
         d      
   m  m         
 d      d       
    d d         
```

Additional aid is also gained by the tube of the corolla of *Teucrium* being resilient; the anterior pair of stamens form two thick ridges, much aiding it in this respect; the posterior pair, however, are, so to say, "sunk" into the tissue of the corolla as to be invisible in a transverse section.

TRANSITIONAL FORMS.—We may sometimes, as it were, catch the formation of irregular and zygomorphic flowers in the process of formation; for it not infrequently happens that one genus will be irregular amongst its allied regular ones. Thus *Verbascum* and *Petunia* are transitional genera, and stand intermediate between *Solanaceæ* and *Scrophularineæ*. The former genus has a less zygomorphic corolla than many of the latter order, and also retains the fifth stamen in varying degrees of utility. We might regard both these genera as Solanaceous, and on the road to acquiring zygomorphism, but to which neither has yet fully attained.

"The short-tube [of *Verbascum nigrum*] widens out into a flat, five-lobed limb, which takes up an almost vertical position; the inferior lobe is the longest, and the two superior are shorter than the lateral lobes, so that an insect settles most conveniently upon the inferior. The stamens project almost horizontally, but curve slightly upwards from the tube, and diverge slightly from one another; they alternate with the petals, and again the superior is the shortest, and the two inferior longer than the lateral ones. . . . The style is shorter than the inferior stamens, and bent down slightly below them"

From this description, taken from Müller's work,* which, with slight modifications, would describe *Petunia* as well, the reader will see how these flowers fulfil the requirements of self-adaptation to insect agency; and in every point of detail are they responding to the forces impinged upon them. The weight of the insect being well to the front, hypertrophy is commencing on the anterior side, while atrophy follows on the others, there being no special thrust as yet on the posterior side of the flowers

There are many other genera and species which stand in intermediate positions between others, and it has always been a matter of doubt to systematists as to which they should be referred. The interpretation of their existence I take to be as here described, namely, that they are in an actual transitional state, brought about by insect agency, if

* *Fertilisation*, etc., p. 429.

they be flowers visited; or by fluctuating conditions of nutrition, if not; and then, arrested in that state.

A further remark on a significant point may be added on *Petunia*. In this flower, as in *Verbascum*, the limb of the corolla stands in a vertical plane, the anterior lobe is a trifle larger than the others, the five stamens have a slight tendency to be atrophied on the posterior side, while the stigma has become just so much displaced as to hinder self-fertilisation. This property is, however, by no means yet lost. Florists are aware of it, and find it necessary to self-fertilise, but not to cross, these flowers artificially to secure plenty of seed; Mr. Darwin corroborates this (*Cross and Self*, etc., p. 193).

We have, then, here a case, but by no means an isolated one, in which the *forms* of the floral organs are undergoing a change, but the *physiological* characters of the essential organs have not yet been influenced by the external stimulus, so as to become more or less inert upon one another, as is sometimes the case in highly differentiated flowers.

Indeed, it would seem to be a universal rule that morphological changes are more readily acquired than physiological barrenness; as by far the great majority of plants have retained their self-fertilising powers; and, when they have lost it, it is easily and rapidly reacquired when the necessary conditions are supplied.

Echium is another instance of almost a single genus amongst others of the same order characterized by great and persistent regularity. *Rhododendron* and *Azalea* may be compared with other genera of *Ericaceæ*, and the reader will readily suggest others.

Sometimes the irregularity is confined to the stamens or style, or both, which may have a tendency to become declinate, as in *Calluna*, in some Liliaceous and Amaryllidaceous

THE ORIGIN OF "ZYGOMORPHISM." 121

plants, as *Narcissus Corbularia*. In *Anagallis arvensis* and *Lycium barbarum* there is nothing but an obliquity in the style observable.

In all the flowers which tend to show irregularities the rule is that the corolla-limb stands in a vertical plane, so that the flowers are visited from the front. This I take, as mentioned above,. to be generally a primary necessity for bringing about irregularities of all kinds. There are some campanulate and pendulous flowers where irregularity occurs in the lengths of the filaments or the size of the anthers. Thus, I have observed great fluctuations in the stamens of *Narcissus cernuus*: some of these I have illustrated in Fig. 37.

I noticed that a peduncle always bore the same form in every flower of its umbel. There were mostly three flowers in each, as of *a*, *b*, and *d*; one specimen of *a* and one of *e* had only a single flower; and one of *c* had two flowers. In *a*, *b*, *c*, *d*, the three short stamens, as well as the three long ones, were all of the same height, respectively; but in *e* one of the shorter set was taller than the rest.

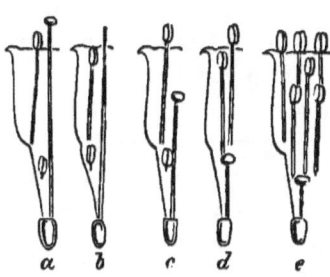

Fig. 37.—*Narcissus cernuus*.

Similar fluctuations are not at all uncommon in cultivated heterostyled plants, as Primroses; as will be alluded to again in discussing the conditions of heterostylism.

In *Fritillaria Meleagris*, though no irregularity occurs in the perianth leaves, it often appears in the androecium, and is more especially observable in the lengths of the anthers. This would seem, therefore, to be another instance of incipient change.

Calluna vulgaris is likewise just, commencing to be

irregular. The flowers are almost horizontal, closely compacted against the axis, and consequently not readily visited on any side except from the front. The style and stamens curve upwards, so that "the smaller bees and flies thrust the head or proboscis from the front into the flower, and the upward curvature of the style and stamens causes the insect to enter by the lower half of the flower, and so to get dusted with pollen from above." *

Müller also notices, about this flower, that "the style, which even in the bud overtops the stamens, grows very markedly after the flower opens, as the flower itself does. As a rule, it attains its full length only after the anthers have completely shed their pollen, at which time also the four-lobed stigma reaches its full development."

He gives five figures of *Saxifraga Seguieri* to show the progressive stages of development. In the first or female (protogynous) condition the stigmas only are mature, the anthers, petals, and sepals being far from having attained their full size. It is not until half the anthers have shed their pollen, and the others ready to do so, that the flower attains its complete dimensions †

I refer to these facts, which are equally applicable to many other flowers, to show that growth normally continues after insects have commenced to visit flowers; so that there is plenty of opportunity for the petals, stamens, etc., to respond to the insect's action before reaching maturity.

Dr. F. Noll has investigated the various movements of zygomorphic flowers during growth, resulting in the external position of the flower; and he finds that the excess of weight on one side is, when necessary, counterbalanced by active tensions (see *Jl. R. Mic. Soc.*, 1887, p. 612 and reffs.).

* *Fertilisation*, etc., p. 379.   † *Ibid*, p. 244.

CHAPTER XIII.

THE EFFECTS OF STRAINS ON STRUCTURES.

VEGETATIVE ORGANS.—In explaining the origin of irregular flowers by insect agency, it will not be amiss to fortify the theory by describing other instances apart from flowers, and to add further results which I believe to accrue from the persistent action of insects on the one hand, and a ready response on the part of the organ on the other.

Researches into the anatomy of stems have proved the existence of this responsive power. Thus, a tree will develop wood in a particular direction if it be compelled to meet special strains imposed upon it; for Andrew Knight found that when trees were allowed freedom in one direction only, and were thus made to oscillate in definite directions, either east and west or north and south, the stem became elliptical in section, the long axis corresponding to the direction of oscillation. Mr. Herbert Spencer has also described how Cactuses, if submitted to particular strains, develop wood to meet them.

The various kinds of the supporting tissues of pedicels, such as collenchyma, sclerenchyma, the so-called liber-fibres as well as true woody fibre, are all so many contrivances of the stems to support the weight of the flowers and fruits, and to overcome gravity. So, again, in the case of apples and pears, if they hang vertically downwards they grow as symmetrically round the insertion of the stalk as an orange; but if the pedicel projects obliquely from the branch, they then thicken along the *upper* side, forming a sort of buttress running down into the stalk, which also itself tends to thicken. This enlargement, which gives the peculiar "lopsidedness" to several kinds of pears especially, and in a lesser degree to some sorts of apples, is simply due to the fact that the force required to counteract the resultant of the two forces, gravity and tension—which act vertically downwards and along the stalk, respectively—must be increased in proportion as the direction of the stalk approaches the horizontal one. The accompanying diagram (Fig. 38) represents the basal end of a Dr. Jules Guyot pear and in the position in which it hangs upon the tree. The letter w (weight) is in the line of gravity, t (tension) acts along the stalk, while r counteracts the resultant, which tends to tear the pear from the stalk at the *upper* side. This strain must be met, and the increased thickness along this upper side enables the pear to resist it, and thus prevents the fruit, especially if it be a large and heavy kind, from being wrenched from the stalk.

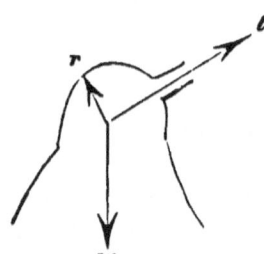

Fig. 38.—Diagram of the end of a Dr. Jules Guyot pear.

A somewhat similar development often occurs with plums and lemons; only, as there is no receptacular tube in either case, the weight of the fruit causes them to produce a thick fold in the carpel on the *under* side, together with some degree of hypertrophy on the *upper*, where the tension occurs.

It is not uninteresting to notice how branches of trees similarly sustain the strain produced by their own weight. This is done by growing at an acute angle (originally caused

THE EFFECTS OF STRAINS ON STRUCTURES. 125

by arising in the axil of a horizontally inserted leaf), much more often than in a strictly horizontal direction. The branch, after growing for a short distance upwards, generally bends downwards, assuming just the same curvature as of declinate stamens which have to support the weight of insects.

If the vertical line in the adjoining diagram (Fig. 39) represent the trunk, and the curved line a branch, the insertion at f supplies the fulcrum, w is the weight of the branch, and acts in a vertical line, p is the power required to counteract the resultant of these two forces.

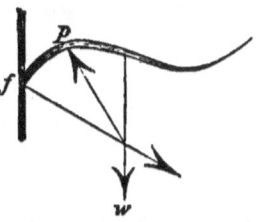

Fig. 39.—Diagram of a tree and branch, illustrating the distribution of forces.

When the bough breaks, either through an additional weight of snow or by its own weight on decay, it snaps off at the point p, *i.e.* the place where the force acts, as it can no longer overcome the resultant of f and w.

REPRODUCTIVE ORGANS.—Applying these principles to floral structures, we have already seen in how many ways the strain to which parts of flowers are subjected, through the weights and pressures of insects, are met and overcome.

In a large number of instances the organ becomes curved, and assumes the character of a spring, yielding on pressure, but recovering its position when pressure is removed. It is often so with the claws of the petals of papilionaceous flowers, the stamens of *Dicentra*, *Corydalis*, and *Veronica Chamædrys*. Similar structures are seen in many styles, as those of Pansy (Fig. 54), and in genera of *Polygalaceæ*.

All declinate stamens partake of it to a more or less degree. The distribution of the forces brought into play to support the insect is exactly the same as when a bough has to support its own weight, as will be easily understood from what has been described, and by referring to the diagram (Fig. 40a).

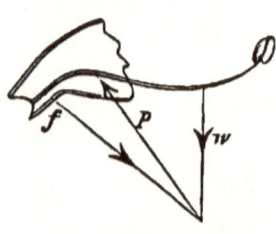

Fig. 40a.—Diagram of declinate stamens, illustrating the distribution of forces.

If the tissue does not remain firm under pressure, then the lever-action of a spring may fail to be secured, and the organ will oscillate freely, as on a pivot. This I take to be another result of a constant, but of course unconscious, effort of the insect to push the organ in a certain direction. It is thus that anthers become versatile, and oscillate, and may become even inverted in position, when pollination is being effected by insects. Consequently anthers normally introrse can be made to assume a pseudo-extrorse position. This happens with some *Cruciferæ* as *Cardamine pratensis*, Tulips, etc. A similar cause I would attribute to the formation of the oscillating anthers of *Salvia*, and of the species of *Calceolaria*, as *C. Pavonii*, which form the section *Aposecos* of that genus, as shown in Fig. 32, *a*, p. 109.

Fig. 40b.—*Lamium album*, showing distribution of forces.

As an example of an entire flower illustrating the distribution of forces, the accompanying figure of *Lamium album* (Fig. 40b) will explain how the forms of the calyx and corolla are adjusted to bear the weight of the insect. The bee alights on the lip and then partially crawls into the expanded mouth of the corolla, so that its weight now lies in the direction of w. The fulcrum will be at f, and the resultant of these is in the opposite direction to r. This is where the strain will be felt; so that it is just at this point where the backward curvature takes place which gives strength to the corolla-tube. This latter is also greatly supported by the tube of the calyx, which, as stated, has a curiously thickened cylinder within the mesophyl.

Finally, if we may admit the existence of this adaptability to strains, and other external forces, and that the various structures of flowers will grow in response to them and develop themselves accordingly, we have a clue to the interpretation of every one of the most diverse forms which may be found in flowers adapted to insect agency.

Similarly, with regard to several classes of cell structure which are now recognized as having a supportive function, such as collenchyma, sclerenchyma, wood fibres, etc., I would contend that such are not formed originally and anteriorly to the requirements of the plant; but that strains have been responded to, and the tissues formed accordingly. Then, subsequently, hereditary influences have come into play, so that *now* they may appear even before there is any actual necessity for them.

I find that M. J. Baranetzki's observations * on the thickening of cell-walls tend to corroborate this view; for he, too, has arrived at the conclusion that the secondary formations on the *interior of the cell-walls* are always in adaptation to protect the cell-wall against the pressures exercised upon it.

In alluding to the above instances of levers and mechanical powers in plants, one mentally recalls how abundant they are in the distribution of the bones and muscles in vertebrates. These latter are, of course, situated only and exactly where they are required. I cannot help thinking, therefore, that the old view was fundamentally correct; that such have been gradually brought into existence by the efforts to meet the strains put upon them. If this be true, then one and the same law has prevailed in the evolution of organs in both the animal and vegetable kingdoms.

* *Ann. des Sci. Nat.* (*Bot.*), iv. (1886) p. 135.

CHAPTER XIV.

ACQUIRED REGULARITY AND "PELORIA."

REVERSIONS TO REGULARITY.—Dr. Masters observes that "in cultivated Pelargoniums, the central flower of the umbel or 'truss' frequently retains its regularity of proportion, so as closely to approximate to the normal condition in the allied genus *Geranium*; this resemblance is rendered greater by the fact that, under such circumstances, the patches of darker colour characteristic of the ordinary flower are completely wanting, the flower being as uniform in colour as in shape. Even the nectary, which is adherent to the upper surface of the pedicel in the normal flower, disappears, sometimes completely, at other times partially. The direction of the stamens and style, and even that of the whole flower, becomes altered from the inclined to the vertical position. In addition to these changes, which are those most commonly met with, the number of the parts of the flower is sometimes augmented, and a tendency to pass from the verticillate to the spiral arrangement manifested." *

All the differentiations in an ordinary lateral blossom of *Pelargonium* brought about by insect agency are, in the above instances, reversed in consequence of the terminal position of the flower. A more complete illustration of the effect of manner of growth and the distribution of nutrition could not

* *Teratology*, p. 221.

well be given, showing how all the features of irregularity acquired by the ordinary form must have been induced or impressed upon the flower when growing laterally and easily visited, but that they are readily lost as soon as the sap can be distributed radially and so cause the parts to grow symmetrically round the now vertical axis.

Besides the occasional appearance of one or more terminal and regular flowers among a truss of irregular ones, it is the object of florists to induce all the blossoms of many irregular flowers to become regular. Thus cultivated Pelargoniums, Gloxinias, Azaleas, Pansies, etc., which are normally irregular, tend to become regular under cultivation, and lose their characteristic features.

In all these cases I am inclined to recognize *negative evidence* in favour of the theory advanced; in that, presuming the characteristic irregularities to have been brought about by the agency of insects and through the crossing of distinct flowers by these creatures, and that the irregularities have arisen under the various pressures, etc.; then, under cultivation, though they may be repeatedly crossed by man—the process, however, not being effected in the same way as by insects, and consequently the causes of irregularity being wanting—the flowers now revert to their ancestral forms; while ample supplies of nutriment doubtless play an important part in the process.

Moreover, though any irregular flower may become regular, it is a significant fact that normally regular flowers are never known to suddenly assume any definite irregular form.

That the change from irregularity to regularity is an acquired constitutional affection is seen in the fact that, when the flowers of a drooping *Gloxinia* are fertilised with their own pollen, a large number of the seedlings will bear the erect regular form of flower.

In the preceding cases the regularity occurring in normally irregular flowers is due to the non-development or arrest of the usually characteristic features which give rise to the irregularity; so that the resulting form is a reversion to, or a restoration of, the ancestral conditions of the flower which is assumed to have been perfectly regular.

As insects, by their mechanical actions, are here believed to have brought about irregularities in flowers; so, conversely, regularity can be reacquired through their agency in another way. *Clerodendron* is a plant in the corollas of which certain members of the family *Tingidæ* take up their abode as pupæ. The irritation induced by their presence brings about a hypertrophy of the corolla, which now assumes a regular form, while the filaments and style are likewise affected, becoming much thicker than in the normal, irregular flower.

Reversions to regularity may, therefore, I think, be safely referred to nutrition as the immediate agent, though such extra flow of nutriment may be brought about by diverse causes.

"Peloria."—Regularity may, however, arise in another way, by the members of the whorl or whorls normally irregular being all exactly alike. Instead of there being any arrest, there is here an excess of development. Thus, if, instead of the anterior petal of *Linaria* being the only one provided with a spur, all the petals become spurred, then the corolla will become regular; but there is no other tendency to revert to the ancestral form. This variety constitutes the form called "Peloria" by Linnæus.

There are, then, two factors, which appear either singly or together, in this process of change. First, a terminal position, as this tends to produce regularity in consequence of an equable flow of sap in all directions: just as this also determines the persistent regularity of all flowers which are normally so situated and are visited from all directions. It will be often found that when Snapdragons have pelorian blossoms they are in three-flowered cymes as in Calceolarias, instead of a raceme, of which the central one is regular, while the lateral flowers are irregular. Secondly, whether terminal or not, the influence which first brought about the change in the anterior part of the flower spreads to and effects all the rest. This statement, of course, only expresses what one sees, without explaining the process; but the fact that the energy peculiar to the formation of one organ can affect others is so common, that we may recognize the process as a principle of growth; just as stamens may become petaloid, on the one hand, or pistiloid on the other; showing that "petaline energy" can affect the andrœcium in the first case, and "pistiline energy" in the latter.

That the true pelorian form is correlated to vegetative energy is seen in the fact that such a flower obviously requires more material than a normal one, and that petalody of the stamens frequently accompanies the modification. Moreover, although of course usually sterile under such circumstances, yet pelorian Linarias have been reproduced when the seeds were sown in a rich soil. Mr. Darwin also raised sixteen seedling plants of a pelorian variety of *Antirrhinum* artificially fertilised by its own pollen, all of which were as perfectly pelorian as the parent plant.

That peloria is due to hypertrophy is also seen in the fact that it always arises by multiplication of the normally enlarged organ. Thus, in *Linaria* and *Antirrhinum* all the petals are spurred or pouched; in pelorian Larkspurs and Aconites it is the spurred and hooded sepal which is repeated; and in papilionaceous flowers it is the standard which is multiplied five times, etc. An abnormal increase in the number of petals and stamens often occurs in pelorian Pelargoniums, Horse-chestnut, etc.

If pelorian forms were equally constant as the one-spurred condition, botanists would undoubtedly have recognized them as species, or perhaps genera, as it is the comparatively slight difference in the length of the spur upon which they separate *Linaria* from *Antirrhinum*. Similarly *Corydalis* has normally but one spur and one nectary. It, however, bears occasionally two spurs and has two nectaries, as in *Dicentra*.

" Peloria, then," as Dr. Masters observes,* "is especially interesting, physiologically as well as morphologically. It is also of value in a systematic point of view, as showing how closely the deviations from the ordinary form of one plant represent the ordinary conditions of another; thus the peloric 'sleeve-like' form of *Calceolaria* resembles the flowers of *Fabiana*, and De Candolle, comparing the peloric flowers of the *Scrophulariaceæ* with those [the normal ones] of *Solanaceæ*, concluded that the former natural order was only an habitual alteration from the type of the latter. Peloric flowers of *Papilionaceæ* in this way are undistinguishable from those of *Rosaceæ*. In like manner we may trace an analogy between the normal one-spurred *Delphinium* and the five-spurred *Aquilegia*, an analogy strengthened by such a case as that of the five-spurred flower of *Delphinium*."

* *Teratology*, p. 236.

CHAPTER XV.

THE ORIGIN OF FLORAL APPENDAGES.

EPIDERMAL TRICHOMES, ETC.—While all conspicuous flowers invite insects of some sort or another to visit them, which, by so doing, pollinate their stigmas, it is an important thing to be able to exclude those which would rifle the flower of its treasures and yet not transfer the pollen from one flower to another. Dr. Kerner, in his interesting work entitled *Flowers and their Unbidden Guests.* has described and figured a large number of instances of the forms of flowers in which he detects various processes, some of which produce sticky secretions, others occurring as hairy " wheels " and " tangles " of wool, etc.; all of which tend to stop the ingress of ants and other small insects, and thus prevent them from getting at the honey. The question at once arises, How have these processes been caused? Without attempting to account for all, the theory I offer will, I maintain, be answerable for a good many, especially for several cases of secretive processes and for the hairy obstructions. All these I would suggest as the immediate results of the irritations set up by insects; so that, as a consequence, they occur just and only where they are wanted; so that, while they form no hindrance to the larger and stronger insects which have presumably caused them to be developed, they, however, may effectually prevent the smaller ones from entering.

In many cases the capability of the flower to restrict itself to its proper visitors, and at the same time to exclude the wrong ones, is a common result of the differentiations which have taken place. Thus, an elongated tube, as in Evening Primrose, and in some species of *Narcissus*, etc., is a direct result of and adaptation to the long proboscides of Lepidoptera, and in proportion as the tube is elongated so does it prevent the ingress of short-tongued insects, or of those with short proboscides.

Apart, however, from such and other general results of adaptations, whereby flowers have become, for example, irregular, and consequently their insect visitors are more and more restricted in number, there are innumerable outgrowths of various kinds which act as special obstructions to the entry of small insects which would not be able to pollinate the flower. Thus, while many regular flowers, such as Gentians, have developed horizontal hairs all round the entrance to the tube of the corolla, Honeysuckle and *Veronica Chamædrys*, which are irregular and approached from one side only, have developed them in the anterior side alone. In *Amaryllis belladonna* Kerner describes and figures (Fig. 41) a one-sided flap growing out of the perianth, and so folded as to furnish a very small orifice for the entrance of a proboscis. There is no such growth on the anterior side, but only on that one, the posterior, which is probed by an insect.

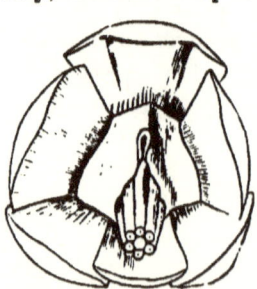

Fig. 41.—Base of flower of *Amaryllis* showing honey-protector (after Kerner).

In *Gentiana Bavarica* there are tooth-like processes at the entrance of the tube, which remind one of the appendages to the corolla of some of the *Sileneæ*. *Monotropa glabra* and *Daphne Blagayana* agree in having a large circular

THE ORIGIN OF FLORAL APPENDAGES. 135

stigma nearly blocking up the tube; and while in the former the irritation set up by the proboscis of an insect has (presumably) given rise to a glutinous secretion, in the latter it has caused a development of hair.*

Did we but know what the insects were, and how they have poised themselves upon the flower, and in what way their proboscides and tongues have irritated the different parts, one might be able to describe more accurately the whole process; but that such has been the cause and effect, as above described, seems to me to be too probable a theory to be hastily discarded in the absence of a better one.

It is one of those arguments of deduction that escape the opportunity of verification, and can only rest for support upon the number of coincidences which can be found, and which collectively furnish a probability of a high order.

When, then, we find that these processes always occur just where we know the heads, legs, bodies, and proboscides or tongues of insects habitually are placed and irritate the flower, we are justified in recognizing, not only a coincidence, but a cause and effect, though we may not be able to trace the action in each individual case. Thus, it may be asked,

* The remarkable fact of Heliotrope being the solitary exception out of the order *Apocynaceæ*, with the stigma forming a circular rim *below* the summit, may meet with its interpretation from a like cause. The corolla is so folded round the style that it leaves no space between it and the latter. Hence it may, perhaps, have been due to a similar "rubbing," that has transferred the stigmatic surface from the now abandoned apex to a lower level, just where the style-arms ought to begin to diverge. The papillæ, too, differ from the ordinary form in being pointed like fine hairs. The relative differences in the distribution of the papillæ on the style-arms of the *Compositæ*, I would also suggest as having been brought about by different insects which irritate them in various ways. So, too, the diverging stigmas of insect-fertilised cruciferous flowers may be compared with the small globular form of self-fertilising species of the *Cruciferæ*.

Why are the three anterior petals of *Tropæolum* fringed, but the two posterior, which stand a long way behind, not so? Why are hairs produced on the anterior side of a Honeysuckle and *Veronica*, but all round the mouth of the regular *Gentiana*? And many other questions of a like sort might be raised. If we watch the habits of insects with their tongues, we may easily see how they irritate the various parts by licking them, not solely where the honey is secreted, but the filaments, etc. Thus Müller often watched Rhingia rostrata licking the staminal hairs of *Verbascum phœniceum*, and in many cases the hairs on the filaments offer a foothold to the insects while visiting the flowers, as in species of Mullein; such hairs, if my theory be true, being the actual result of the insects clutching the filaments or rubbing them with their claws. In *Centaurea*, the epidermal cells of the filaments have produced projecting processes just where the proboscis rubs against them when searching for honey in the little cup (see Fig 11, p 60), from the middle of which the style issues, as shown by the direction of the arrow.

These filaments also exhibit their extreme irritability by contracting, and so assisting in the "piston action" by dragging the anther-cylinder downwards over the style.

While recognizing the coincidence between the localization of outgrowths, enations, trichomes, etc., and the position of the parts of insects in contact with flowers when searching for honey, one must not forget that a great number occur where such contacts do not take place. Hence we must look for other possible causes for their origin as well. One of the commonest forms of trichomes is glandular hairs, and, as Dr. Kerner has pointed out, when they occur on sepals, pedicels, etc., they form admirable barriers to the approach of ants and other creeping insects, which might rifle the flower and yet not fertilise it. We must be on our guard, however, in

THE ORIGIN OF FLORAL APPENDAGES. 137

asserting that nature has produced them *in order* to keep ants off; for that line of reasoning is pretty sure to land us in faulty teleological methods. What causes them is not at present known in all cases; though we may perceive that certain conditions, as growth in water, can bring about their disappearance, as Dr. Kerner remarked in the case of *Polygonum amphibium*, which only has them when growing on land.

If, however, we ask, for example, why the Sweet-briar has them all over it, and why the Dog-rose has none, I do not know how to reply to the question as yet. We may notice certain coincidences, that hairy herbaceous plants are commoner in dry situations and smooth ones in watery; just as root-hairs occur in a loose sandy soil and their absence is noticeable in a heavy one; but we do not know how these different media actually bring about these changes, though we may feel assured that it is solely due to the environment.

If we, thus, look elsewhere than in flowers for any analogous processes they are by no means wanting. For example, it is simply the mechanical irritation brought about by contact with a foreign body, probably aided by moisture and a lessened degree of light, that causes the epidermal cells of the aërial roots of the Ivy and Orchids (Fig. 42) to elongate into adhesive or clasping hairs, so as to grasp the body for support. This is only a form of the ordinary root-hairs which are immediately developed when the tip is in contact with a moist soil, and each hair grips and glues itself

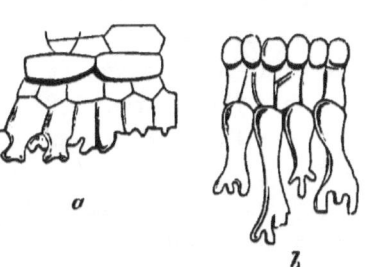

Fig. 42.—Adhesive epidermal cells of roots of Orchids: *a*, aërial; *b*, subterranean (after Janczewski).

to the particles of soil.* Chatin noticed the production of hairs when the roots came in contact with any obstacle; † but Dr. M. T. Masters observes that the obstacle alone in their case is insufficient without moisture, for he found that the roots of Mustard-seed could penetrate a stiff clay, but did not develop any root-hairs until they came in contact with the sides of the pot—" Wherever there was a thin film of water investing a stone or the sides of a porous flower-pot or a plate of glass, there the root-hairs abounded."

Besides a nutrient or moist medium, actual growth in water may enormously increase the length and quantity of root-hairs; as may be seen in the dependent roots of floating plants of *Hydrocharis*, etc.; or in the hypertrophied conditions of the roots of grasses when growing in water.

That epidermal trichomes may be due to the irritation of insects is clearly seen by their appearance within the cavities of certain galls. ‡ In the case, for example, of a very common one on willows, the leaf bulges out below and forms a sort of bag, open or closed above. The tissues become hypertrophied though the epidermis and palisade cells are still recognizable lining the cavity. The leaf has scattered hairs on both sides; but within the cavity much larger hairs, rich with protoplasmic or other matters, project from all sides into the interior. Some are straight, others curved, club-shaped, or with irregularly swollen ends, not unlike the forms produced on climbing roots by contact with a foreign body. Again, the crimson "spangles," so common on the underside of Oak-leaves, are covered with stellate clusters

---

* Sachs' *Phys. of Pl.* (Eng. ed.), 1887, fig. 12, p. 19.

† Mem. Soc. Nat. Sci., Cherbourg, 1856, p. 5; referred to by Dr. M. T. Masters in *Notes on Root-hairs*, etc., Journ. Roy. Hort. Soc., vol. v., p. 174.

‡ Caused by species of Nematus.

of hairs. Similarly, those of *Cecidomyia Ulmariæ* on *Spiræa Ulmaria* are hairy outside, and papillose within; while similar ones of a Phytoptus on the Sycamore are lined with long blunt-ended hairs, and are clothed without by others, long and pointed. In all these cases the galls, as well as the hairs, are the product of irritation set up by the presence of the egg deposited by the insect.*

As another very common instance of the presence of epidermal papillæ and hairs, may be mentioned their occurrence in the stylar and ovarian cavities. The former, and the placentas especially, may be clothed with delicate hairs exactly resembling root-hairs. Such may be well seen in the Poplar, *Tamus, Richardia Æthiopica*, etc.; and since M. Guignard † has discovered that the mechanical and physiological irritation of the pollen-tubes is required to cause their development on the walls of the ovary in *Vanilla*, between the longitudinal bands of conducting tissue, it is, I think, a by no means improbable theory that the tufts of hairs over the nectaries, "tangles," "wheels," etc., on the filaments or corolla-tubes, have been actually caused by the irritation of insects, since they occur just where such irritations are made.

One use of certain outgrowths has been regarded as intended to protect the honey from rain. Why, however, some flowers should be so favoured while many others, as of the *Umbelliferæ*, have no protection at all, is not stated. The interpretation I have here offered will, of course, apply to all such growths, whenever they may really keep off rain or "unwelcome guests."

* Krasan has lately discussed the formation of the woolliness of galls, etc., *Oesterr. Bot. Zeitschr.*, xxxvii. (1887), pp. 7, 47, 93, seqq.

† *Sur la Pollinisation et ses Effets chez les Orchidées*, par M. L. Guignard, Ann. des Sci. Nat., tom. iv., 1886, p. 202.

CHAPTER XVI.

SECRETIVE TISSUES.

POSITION OF NECTARIES.*—These honey-secreting organs seem capable of being formed anywhere. Of course they are mainly to be found in flowers, but many plants bear them elsewhere. Thus, some ferns have them on the rachis; the common laurel, as also the almond and peach, have two at the base of the petiole; beans and vetches, as well as species of *Impatiens*, have them on the stipules, as shown in Fig. 43. Bees may be often seen as busy about the young shoots of laurel as if they were visiting flowers. *Acacia sphærocephala* has a large one, on the upper side of the petiole, which supplies those ants with food which take up their abode in the gigantic stipules peculiar to that genus.†

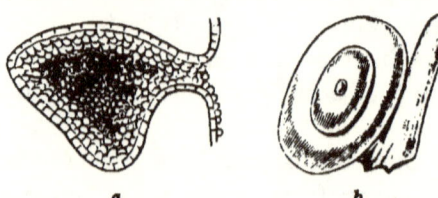

Fig. 43.—Stipules of *Impatiens*: a section showing anatomy; b, with a drop of honey in the centre (after Kerner).

* *Les Nectaires*, Ann. des Sci. Nat., Bot., vol. iii., p. 1, 1879; also, *Études Anatomiques et Physiologiques des Nectaires*, Compt. rend., tom. lxxxviii., p. 662, 1879; also, *Cross and Self Fertilisation of Plants*, p. 402; also, Stadler, *Beitr. z. Kenntniss d. Nectarieen u. Biologie d. Blüthen.*

† See Belt's *Naturalist in Nicaragua*; also a paper by F. Darwin, in Trans. Lin. Soc., on the same subject.

SECRETIVE TISSUES.

A microscopic examination of the anatomy of nectaries shows them to be composed of small cells closely resembling the merismatic condition of ordinary cellular tissue (see Fig. 43, a), and similar to the arrested parenchyma of the pulvinus at the base of the petiole of sleeping leaves, which enables that organ to remain flexible. Or, again, it is very similar to the conducting tissue of the style, which owes its origin to the irritating effect of the pollen-tubes (chap. xviii.).

The function of the nectary is to secrete honey, or, to speak more accurately, either principally glucose, or else cane sugar, or both, for the proportion varies greatly.*

The position of nectaries in flowers is very various, and any organs can form them. It will be enough to enumerate a few localities as follows: The Lime, species of *Malpighia*,† and perhaps *Coronilla*, furnish instances, which are comparatively rare, of the sepals of the calyx being nectariferous. In Buttercups, Hellebore, and Aconite, nectar is secreted by the petals or their representatives. In Violets, *Atragene* (Fig. 44), *Pentstemon*, and *Stellaria* the filaments undertake the duty, while in *Caltha*, *Monotropa*, and *Rhododendron* it is the carpels or pistil. In most instances the honey is secreted by glands, disks, etc., issuing out of the floral receptacle. If the ovary be inferior, then the secreting structure is on its summit, as in the *Umbelliferæ*; and in that case it is the base of the styles from which the nectariferous tissue is developed.

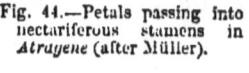

Fig. 44.—Petals passing into nectariferous stamens in *Atragene* (after Müller).

THE ORIGIN OF NECTARIES.—Limiting one's self to those in

* Bull. Soc. Bot. Fr., viii. (1886), *Rev. Bibl.*, p. 212.
† *Nature*, vol. xvii., p. 78.

flowers, there are many reasons for inferring their existence to be due to the direct and irritating action of insects themselves when searching for juices as food or otherwise.

That a merely mechanical irritation may cause a flow of nutrient fluid to the spot, so that the tissues may increase in size by the development of cells, which would not otherwise occur, is abundantly evident. It is seen, for example, in the growth and development of galls; of the so-called "Ant-plants" on *Myrmecodia* (p. 115), *Acacia sphærocephala*, etc.; in the thickening of all climbing organs as soon as the irritation of the foreign body has commenced; hence the inference that hypertrophy may occur wherever an insect's proboscis can irritate the floral organs, is by no means without foundation. Why the cell-contents of nectaries should especially give rise to sugar, is a question at present beyond answering. Those of conducting tissues appear to do the same. In the case of nectaries it may, perhaps, have originated as a pathological phenomenon which has become fixed and hereditary; for pathological conditions often determine a flow of gum, as in Cherry-trees, resins in the *Coniferæ*, watery and sugary discharges from wounds, etc.; and it is impossible to draw any hard-and-fast line between a pathological and varietal state: as, for example, in closing the scar after the fall of the leaf the fibro-vascular bundles are sometimes stopped by gum— a process which, in this case, might be regarded as normal, and not pathological as in the former.

If a particular locality be perpetually irritated, so to say, for generations, all analogy shows that the effect may become permanent and hereditary; at least, as long as the irritation is persistently renewed year after year. And, on the contrary, the theory is equally supported by the negative evidence of the disappearance of the honey-glands whenever the whole flower degenerates and becomes regularly self-fertilising or else anemophilous. In these cases, in unison with the degradation in size and colour of the corolla, or else its entire loss, the nectaries tend to and generally vanish entirely; as may be seen in *Polygonum aviculare* as compared with *P. Fagopyrum* and *P. Bistorta*.

The simple origin of nectaries, then, according to my theory, is that insects, having been attracted to the juicy tissues of flowers, by perpetually withdrawing fluids have thereby kept up a flow of the secretion which has become hereditary, while the irritated spot has developed into a glandular secreting organ.* These spots occur wherever the prevailing insect found it most convenient to search; hence it is sometimes at one place, sometimes at another, even in closely allied plants. Thus, in Buttercups the stamens and carpels form a compact globe, especially the latter, and defy the penetration of a proboscis. The corolla, however, admits of an entrance of its base. In *Atragene alpina* the basal portion of the filament forms a nectary (Fig. 44). Comparing these with *Caltha*, the large carpels of this plant admit the passage of a proboscis between them; and the nectaries are now developed on the sides of the ovaries, exactly where they would be irritated.

In *Ranunculus cortuscæfolius*, of the Canary Islands, which has a corolla more than two inches in diameter, the petals are entirely without honey-glands. On the other hand, the carpels are very large and flat, with plenty of space between them. Although I could detect no honey in plants grown at Floore, Weedon, the tissue over the centre of the ovary was modified, and exactly resembled the ordinary tissue of a honey-gland. If I am justified in assuming the carpels as

* It is closely analogous to the action of the pollen-tube, which causes a flow of nutriment to the conducting tissue, only there is a physiological as well as mechanical irritation in that case.

nectariferous, this would bear out the above remarks, for it would be as easily accessible as in the case of *Caltha*.

The merely occasional puncture and lesion caused by an insect which then flies away and does not keep up the irritation—unless it be renewed by other insects—would not of itself be hereditary.* Thus, for example, *Anemone nemorosa* appears to be honeyless, but supplies pollen to bees; yet Müller noticed them frequently probing between the sepals and stamens, apparently to obtain juices wherewith to moisten the pollen. This process may have been the actual origin of nectaries, the result of a wound constantly repeated and kept up, being a flow of a sweet secretion, which has thus attracted insects and induced them to repeat the process.

ANALOGOUS CASES.—A somewhat analogous illustration is that of galls, but in them the presence of the egg, and subsequently the grub, keeps up the irritation. These remarkable structures do not form spontaneously as nectaries now do, without a puncture; still, even in this case, there may be, for all we know, a *predisposition* to form them; perhaps seen in the readiness with which the Oak forms so many kinds, and they may be *now*, perhaps, much larger than they were when insects of any particular species first punctured the ancestral oak upon which so many kinds have now been evolved.† The apex of a shoot of Yew attacked by Cecidomyia taxi is transformed into a fleshy ring curiously resembling the honey-disk of many flowers.

It is well known that in the human subject there may be a predisposition for tumorous or cancerous growths which is hereditary; and there would seem to be a very close

* Injuries, especially to the nerves, may be hereditary in man; see *Nature*, xxiv., p. 257.

† M. E. Heckel thinks that the female "gall-flowers" of the Fig, with an abortive ovary, in which the Cynips blastophaga lays its egg, is now an hereditary form (*Bull. Soc. Bot. de Fr.*, 1886, p. 41).

resemblance between tumours and galls, though originating from different sources, both being hypertrophied conditions of certain normal tissues. For example, Sir B. C. Brodie thus describes a fatty tumour: "There is no distinct boundary to it, and you cannot say where the natural adipose structure ends and the morbid growth begins." It is precisely similar with galls, which are due to cell-division setting in at certain points of the epidermis and subjacent tissues.

Although lesions and mutilations will not as a rule prove to have any hereditary effects, yet the tendency to respond to an irritation becomes permanent, and the form and structure of the resulting organ may actually appear long before the irritation is applied. This is conspicuously the case in the tendrils of *Ampelopsis Veitchii*, in which the adhesive "pads" are in preparation before any contact with a wall has taken place. This is not the case with *A. hederacea*. Similarly the aërial and climbing roots of Ivy are regularly produced only on the shaded side. They can, however, be readily made to form on the opposite side, if that be artificially shaded; and where, indeed, they may be not infrequently found in nature, where they can be of no use. Such cases prove that the tendency to produce them is an hereditary affection which is present *before* the irritation is brought into play. Again, with regard to the tendrils of the *Cucurbitaceæ*, though the coiling does not take place till the irritating effect induced by contact with a foreign body has brought it about, yet the tendency is seemingly so strongly hereditary that, in several cases, the tendrils are coiled while undeveloped in the bud, and have to straighten themselves before again coiling on contact, as may be seen in the common Bryony.

In the case, however, of a mutilation, when it has been once made, the place heals over, and there is an end of all special vital action at the place. If, however, the same place be induced to secrete by constantly repeated irritations, as the same flower is repeatedly visited over and over again before it fades, and the flowers of its offspring have to undergo the same process, year after year, generation after generation, I think it is at least a reasonable surmise that there will at last ensue a permanent flow of fluid to the place, with a corresponding modification of structure, and so the nectary becomes established. If, however, from any cause the flowers become neglected, then the nectaries degenerate and ultimately disappear.

Apart from some general theory of the kind proposed, it is impossible to assign a reason for glands appearing at all sorts of places in flowers. A theory to be worthy of acceptance must meet all cases, if possible, and I maintain that the one I propose is compatible with every observation that has been made in flowers.*

---

* I would suggest a similar origin for the insectivorous pitchers of *Nepenthes*. They originate, as Sir J. D. Hooker has shown, from water-glands. The effort to dispose of water brought up by the fibro-vascular cord keeps the tissue of water-glands at the extremity of a cord in a state of plethora, thereby somewhat arresting any change of form and retaining the cells in the very characteristic merismatic stage. And if it now meet with an external irritation from insects attracted by the escape of fluids a further response to their influence begins, and the wonderful structures we are familiar with in the pitchers of *Nepenthes* are the final result.

I see no greater difficulty in conceiving of such an origin than in any other complex structure, such as the human eye. If the latter could originate from an epidermal cell sensitive to light only, and by successive increments, traceable more or less distinctly through the various strata of animal life, finally reach the highest and most complex form of that of man, there is nothing inconceivable in the growth and differentiation of a pitcher in response to an external stimulus.

What I cannot conceive of is, that any organ has ever originated without a definite stimulating cause acting persistently in one and the

SECRETIVE TISSUES. 147

With reference to the continuous flow of nectar, I would draw some analogy from animal secretions. Mr. Darwin, in speaking of the cow, observes * : " We may attribute the excellence of our cows and of certain goats, partly to the continued selection of the best milking animals, and *partly to the inherited effect of the increased action, through man's art, of the secreting glands.*" This fact, recorded in the last sentence, which I have italicized, is only one example of the general principle of increase of growth by use, which I take to be strictly analogous to what takes place in the vegetable kingdom. And we may notice, in its special application to the formation of glands and other structures by mechanical irritation, that it is none other than a mechanical irritation which keeps up the secretion of milk for prolonged periods.

The common or physical basis of vegetable life, namely protoplasm, is very nearly † indistinguishable in its properties from that of animals. Their behaviour is every day being proved to be not only similar but identical in the two kingdoms. The effects, under mechanical irritations and strains, of nutritive matters of the same kind, of poisonous substances, of electricity, etc., all show that the bond which unites the animal and vegetable kingdoms together is of one and the same nature, and that the links of the chain are forged out of this common basis of life.

It is not to be wondered at, then, but rather to be antici-

same direction. In the case of the eye, I take that cause to be light. In the case of an irregular corolla or the pitcher of *Nepenthes*, I assume it to be insects (*Tr. Lin. Soc.*, xxii., p. 415; *Ann. Sci. Nat.*, 4 sér., xii., p. 222).

Conversely, in the absence of light the eye vanishes; in the absence of insects, corolla, honey, etc., go; so that negative evidence tends to support the positive in all cases alike; see *Or. of Sp.*, 6th ed., p. 110.

* *Anim. and Pl. under Dom.*, ii., p. 300.
† See *Journ. Roy. Micr. Soc.* 1887, 771.

pated, that tissues will behave alike in both kingdoms; that organs will grow with use and degenerate with disuse; that they will develop processes to meet strains put upon them, as the limbs of animals have done and as stems * will do by forming special tissues; and, on the other hand, that they will atrophy if not called upon to display their powers, as parasitic organisms abundantly show in both kingdoms; and as plants degenerate in water, which saves them the trouble of supporting themselves.

All this is exactly what one finds to be the case in every department of the animal and vegetable kingdoms alike, whenever we search diligently into the anatomy and meaning of the histological details of all parts of organisms.

CORRELATIONS OF FLORAL NECTARIES WITH POLLINATION.—There is yet another point observable in glands. As the position of a gland or nectary is just where it is most easily accessible to the particular insects which visit the flower—a fact abundantly illustrated throughout the floral world,—and since the sole use of it to the plant, as far as we can see, is that it should attract insects which transfer the pollen from one flower to another, one naturally looks to see if the positions of the anthers and stigmas are in any way correlated to that of the honey-gland. Such is, in fact,

* I would throw out a suggestion that the anomalous stems of climbers, which often develop supernumerary collateral axes, but all coherent in one common stem, may be due to a response to the strains to which these stems are subjected, occurring in various directions, as they hang dependent on other trees. Other peculiar features, as of innumerable vessels, feeble wood tissues, etc., I take to be due to degeneracy, through these stems not being self-supporting, so that they have assumed very much the anatomical characters of subterranean roots. Again, just as the pericycle plays so important a part in the structure of many roots, it will be found that this same active layer is the parent of at least several of the above-mentioned supernumerary tissues in climbers, as in the tendrils of *Cucurbita*, *Bryonia*, etc.

invariably the case; so that one cannot but infer that a common cause has brought about their correlated positions. This close correlation is, of course, especially observable in the more highly differentiated flowers. In regular flowers, accessible on all sides, the glands are placed symmetrically round the flower—whether on the sepals, as in Lime; on the petals, as in a Buttercup; or on the receptacle, as in *Geranium pratense*,—or else there is formed a disk, as in so many "discifloral" plants. As soon, however, as a flower begins to show some tendency to irregularity, or the flower is visited in one way only, the honey-secreting organ at once becomes more restricted in localization; as in the Wallflower, where it forms two cushions, out of the middle of which the shorter stamens arise, while the petals form two pseudo-tubes leading down to those two glands. Again, in the *Labiatæ*, so markedly zygomorphic, the honey-gland is often restricted to the anterior side, on which the proboscis is inserted. Similarly in *Antirrhinum majus*, "the honey is secreted by the smooth green fleshy base of the ovary, whose upper part is paler in colour and covered with fine hairs; . . . it remains adherent to the nectary and to the base of the anterior stamens. The short wide spur permits the insect's proboscis to reach the honey from below; above and in front it is protected by a thick fringe of stiff knobbed hairs on the angles of the anterior stamens."*

It is hardly worth while giving other cases to prove the universal rule, that the position of the honey and its gland is always where it is most accessible; and the position of the anthers is, at the same time, just where they will be most conveniently struck by the insect; while the style and stigma supply a third correlation, so that the latter organ invariably hits the insect where the pollen has been previously placed.

* Müller, *Fertilisation*, etc., p. 433.

One more point may be noticed in connection with the above-mentioned correlations, namely, the motility of many stamens. This is always in reference to fertilisation, and, if it be an adaptation to intercrossing, then the anther takes up such a position that the insect strikes it when searching for honey, as in the Aconite and *Tropæolum*. If, on the contrary, the motion is to secure self-fertilisation, then it is regardless of the honey, and may actually interfere with the access to it by insects, as in the *Rosaceæ*: for in members of this order, with an indefinite number of stamens, the further they spread away from the pistil the more readily is the honey accessible; but when they curve inwards, and crowd over the stigmas in the centre, they completely cover up and conceal the honey-disk.

The position of the anthers in relation to the honey-secreting organs will, I think, often be found to be the clue to certain anomalies in flowers. Thus in *Geranium pratense* it has been noticed that the petaline stamens stand ultimately externally to the calycine. Now, the position of the five glands in front of the sepals requires that a tubular space should exist above them, down which an insect may thrust its proboscis, as in the Wallflower. Consequently the five stamens in front of the sepals must be so disposed as not to interfere with this passage. This can only be secured by their bending well inwards towards the styles below, and then outwards, above, so as to bring the anthers again on the same vertical plane as those of the petaline stamens.

The more internal position of the calycine stamens, and the external position of the petaline ones, are immediately due to the gland, so to say, forcing the former inwards, while the buttress-like bases of the carpels thrust the latter outwards. This gives rise to the so-called obdiplostemony of the *Geraniaceæ*.

CHAPTER XVII.

SENSITIVENESS AND IRRITABILITY OF PLANT ORGANS.

GENERAL ILLUSTRATIONS—PROTOPLASMIC IRRITABILITY.—Having now stated on what grounds I believe that the cohesions and adhesions between them, as well as the forms of floral structures have arisen—namely, in response to the irritations set up mainly by insect agencies, coupled with the effects of nutrition, atrophy, hereditary influences, etc.,—it will be desirable to show briefly, not only how remarkably sensitive almost all parts, both vegetative and reproductive, are to the action of stimuli, but how they exhibit even visibly responsive effects, both in the protoplasm of the cells and in the tissues which are composed of them.

The sensitiveness of living protoplasm is one of its most marked and well-known phenomena. It exhibits changes in its distribution within the cell as well as motions, which are the direct result of external stimuli. These may be very various, such as light, heat, electricity, or a merely mechanical irritation, as well as organic and inorganic solutions.

Of the effects of stimuli upon the protoplasm, some may be beneficial, and partake of the nature of nutrition, as may be witnessed in the protoplasmic "aggregation" of insectivorous plants.* Very similar appearances follow electrical

* See Darwin's *Insectivorous Plants*, fig. 7, p. 40.

or mechanical irritations. Thus Fig. 45* shows the effect of electrical action on the threads of protoplasm; *a* represents a cell of a hair of *Tradescantia Virginiaca*; *b* the same, after the application of an electrical current. The following are Dr. Weiss's observations upon this phenomenon :—

"A constant electrical current is without influence upon the protoplasmic excitation; whereas the alternate *shocks*†

Fig. 45.—Cell of hair of *Tradescantia*: *a*, normal condition; *b*, under electrical action (after Weiss).

* From Weiss's *Anatomie der Pflanzen*, p. 95.
† Pfeffer has noticed that the weight, *per se*, of the body in contact [if very slight ?] is of no consequence to tendrils. Thus cotton-wool weighing ·00025 grain produced no effect if carefully placed on them; but it did when a gentle *impact* was caused by slight currents of air. Tentacles of *Drosera* have a sensitiveness very similar to that of tendrils, inasmuch as small splinters of glass only produced irritation of the glands when they caused a rubbing as the result of concussion (see *Journ. Roy. Micr. Soc.*, 1886, p. 285).

Pfeffer concludes that the conduction of sensitiveness is not altogether due to a continuity of protoplasm, as it does not extend to the epidermis. Since, however, the outer cell-wall of the epidermis can grow when in contact with a foreign body, it would seem to clearly indicate that under such circumstances it still retained its protoplasm; and that the modern view of the cell-wall being at first a protoplasmic layer, and not altogether a dead secretion from the protoplasm within it, is correct; for otherwise it is difficult to imagine how it could adapt itself to the surfaces of foreign objects at all.

Heckel, in studying the movements of the stamens in *Sparmannia*, *Cistus*, and *Helianthemum*, discovered that the epidermis plays an important part: " L'épiderme, contrairement à ce que voulait Morren (*Ann. des Sci. Nat.*, t. xix., p. 104), est donc dans quelques cas l'organe principal et visible du mouvement. Je me suis mieux assuré du rôle qu'il remplit, en enlevant cet épiderme quand les dimensions des filets mobiles le permettaient sans mutilation profonde (*Cistus ladaniferus*) : tout mouvement alors était suspendu " (*Bull. de la Soc. Bot. de Fr.*, tom. xxi., 1874, p. 212). See below, p. 163.

SENSITIVENESS AND IRRITABILITY OF PLANT ORGANS. 153

of an inductional apparatus always produce more or less deeply extending changes in the form of the plasm, which resumes its normal character if the power exerted be not too strong. The protoplasm immediately forms itself under the inductional shocks into lumps or balls; and, moreover, often sends club-shaped extensions with great suddenness and energy into the cell lumen, and immediately brings the circulation to a standstill. The rotation returns, however, after a period of rest, the extensions are drawn in, and the former net-shaped distribution of the protoplasm is restored, even when the whole mass of the plasm has been changed into a number of colourless balls and lumps. If the current is allowed to go only through a limited portion of the cell, then the streaming stops also in this tract only, and that, too, amid the formation of the lumps and balls.

"A sudden increase and decrease of temperature acts in the same way; there ensues a formation of drops, a cessation of the current, etc. Yet even here a return to the normal constitution takes place if no real coagulation of the protoplasm has occurred. On the contrary, the current often ensues, after its recommencement, with a greatly heightened speed, and even boisterously. The grains, etc., found in the cell-sap outside the protoplasm are, however violently the current may flow, in no way influenced by it, but remain at rest."

M. E. Heckel has described* the effect of a mechanical irritation on the protoplasm of the cells of the filaments of *Berberis*. He says that the cells of the irritable part are arranged in a parallel manner, being longer than broad. Their contents are of a yellow colour, and disseminated throughout the whole cavity, but especially applied upon the walls. After receiving the excitation, the same cells, the surface of which is striated transversely, are massed

* *Bull. de la Soc. Bot. de Fr.*, tom. xxi., 1874, p. 208.

together or aggregated, so as to occupy only two-thirds of the space they formerly required. The contents, retreating from different points of the circumference, are condensed in the centre of the cell, and the transverse striæ are pronounced in a high degree. The cells at the back of the filament are contracted in repose, and extended under irritation, *i.e.* in an opposite manner to that of the other side of the filament. The irritability does not reside in the epidermis.

A result of this aggregation must be a frequent displacement of the nucleus. In Weiss's figure the irritation happened to be made apparently at one end of the cell, while the nucleus was at the other; but in Heckel's description it appears that the protoplasm is drawn from every point; so that, supposing the nucleus had been at the lower end of the cell (Fig. 45, *a*), it would have been most probably displaced. The consequence would be, that if such a nucleus formed its cell-plate, the ultimate position of that plate would be different from what it would have been had no irritation been applied to the organ.

Though one does not look to electricity as a cause in nature, yet that light determines the direction of cell-division is abundantly proved in the case of leaves, whose tissues alter according to their position; the palisade cells, for instance, bring formed on both sides, if the exposure to light be equal, or on the *under side* if that be placed *uppermost*. Similarly does it influence the formation of stomata.* Again, Stahl has shown that the direction of the division of the nucleus, which takes place in the spores of *Equisetum* depends upon the direction of the rays of light; *the two daughter-nuclei lying in the direction of the ray*. On the other hand, the nucleus at the greater distance from the source of light is that of the root-cell, while the one nearer to the source of

* M. L. Dufour., *Ann. Sci. Nat.*, tom. 50 (1887), p. 311. See below, p. 173.

light is that of the prothallium cell.* Climbing roots of Ivy also appear on the darker side of the shoot, etc.

It is impossible to regard the above cases as isolated, but they are special instances, revealing not only the general irritability of protoplasm, but the minuter effects upon the nucleus, which, in its turn, is thus compelled " to respond," and sets up cell-division, *i.e.* the formation of a tissue in the direction of the external influence, as mentioned above in the sentence I have italicized.

The next very important point to notice is that cell-division can take place in response to, and in the direction of an external *mechanical* stimulus, just as well as in that of light. As the sensitive plant is influenced by, and visibly moves its foliage under the irritation of a touch or of varying degrees of light, so do I assume that the peculiar anatomical structures which permit of those motions are the direct result of external stimuli. *Sparmannia*, it may be added, exhibits three kinds of movement, viz., *Sleep* in the calyx and corolla, *mechanical irritability* in the stamens, and an *elevation* of the peduncle. (See Heckel, *l.c.*, p. 210.) If this position be granted we have at least a working hypothesis for the present theory of the origin of floral structures.

FORMATION OF TISSUES DUE TO IRRITABILITY.—Apart from the preceding theoretical supposition, there may be frequently witnessed an actual formation of tissues of various kinds, through hypertrophy on the one hand, often coupled with atrophy on the other, and entirely brought about by physical or mechanical irritations. Cell-division is thus set up, a result which would not have occurred had not the external stimulus been applied.

It is an important fact to notice, that in some cases the abnormal growth, though immediately following the stimulus,

* See *Jl. Roy. Micr. Soc.*, 1886, p. 287; and *Bull. Soc. Bot. Fr.*, 21, p. 65.

and never occurring without it, leaves no hereditary effect as in the case of galls * and of the thickening of the tissues of some climbers after they have caught and clung to a foreign body, such as the petioles of *Clematis*,† and the hooked peduncles of *Uncaria* (Fig. 46). In other cases the effect has become hereditary, and may then be regarded as a specific character. These differences are well seen in the tendrils of *Ampelopsis hederacea* as compared with those of *A. Veitchii*. In the former there are no traces of the adhesive "pads" at the terminations of the slender hooked tips of the branching tendrils, until contact with the surface of a wall has occurred. On the latter species, however, the pads are in course of development before any contact has taken place just as the aërial roots of Ivy begin to appear before contact. It is therefore reasonable to conclude that the effect of contact has become more or less hereditary in the latter Japanese species, though not in the American.

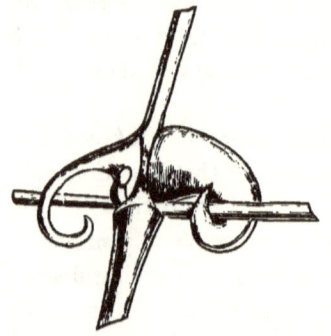

Fig. 46.—Climbing peduncle of *Uncaria*, thickened after catching a support (after Treub).

These tendrils behave exactly like the clasping roots of Orchids, Ivy, etc., as well as the so-called "roots" of *Laminaria, Cutleria*, etc. Indeed, the way in which subterranean root-hairs fix themselves to particles of the soil is by essentially the same method. The irritation caused by contact aided by moisture excites the cell-wall to grow out into protuberant processes, which enables it to adapt itself to the

* I have examined a considerable number of galls, and can quite corroborate M. Prillieux, who has shown how the normal tissues become hypertrophied (*Ann. des Sci. Nat.*, sér. 6, tom. ii. (1876), p. 113).

† See *Climbing Plants*, fig. 1, p. 47, and fig. 4, p. 74.

irregularities of the surface of the particles. An excretion of mucilage appears to follow, which fixes the organ to the foreign support. The irritation not only affects the epidermal layer, but the subjacent tissues as well, which then assist the former in grasping the support.*

Another result of growth due to external agencies is seen in the hypertrophied stipules of *Acacia sphœrocephala* and the stems of *Myrmecodium*, etc., in consequence of the irritation set up by ants. Dr. Beccari † (and M. Treub ‡) has examined these "Ant-plants," which occur in *Rubiaceæ*, *Myristicaceæ*, *Euphorbiaceæ*, *Verbenaceæ*, *Melastomaceæ*, and *Palmæ*, and explains the abnormal structures by variability and heredity. A small swelling appears on the tigellum of *Myrmecodium* serving the purpose of a reservoir of water, but which only grows larger through the agency of ants. These creatures induce hypertrophy of the cellular tissue. This, then, becomes hereditary. I would venture to go further, and attribute the large honey-pits at the base of the leaf-stalk on *Acacia sphœrocephala*, as well as the terminal "fruit-bodies" occurring on the tips of the leaflets, to the same cause, viz. the mechanical irritation of the ants.

There is, in fact, an abundance of evidence to prove that many organs of a plant, if subjected to irritation, can respond to it, and not only increase in size by hypertrophy, but materially alter their anatomical structure and develop new processes. Secondly, that these altered states, if the irritation be persisted in, may become hereditary.

* See Fig. 42, *a*, (p. 137), which represents the inferior side of an aërial root of *Phalœnopsis amabilis* in contact with a surface; *b* is that of a root which has penetrated the soil (*Organisation dorsiventrale dans les Racines des Orchidées*, par M. E. Janczewski. *Ann. des Sci. Nat.*, sér. 7, tom. ii., p. 55.

† Malesia, ii. (1884). See *Arch. Ital. de Biol.*, vi. (1885), p. 305.

‡ *Ann. Jard. Bot. Buit.*, iii., p. 129 (1882).

THE STRUCTURE OF FLOWERS.

The influence of the environment upon the anatomical and morphological structures of plants has been lately and widely studied from several points of view; and it has been shown conclusively, by Constatin and others, how a change of medium—as, for example, from air to a subterranean one, or, again, to water—profoundly affects every tissue of the plant, whether the root, stem, or leaves be submitted to it. So, too, leaves of many plants have been proved to be very sensitive to changes of position and to different amounts of light—which is a most potent and exciting cause in affecting the mesophyl, palisade, and other tissues, including the epidermis, stomata, and even cuticle. It is foreign to my purpose to describe or discuss these details in the vegetative system of plants; my sole object being to draw attention to the fact, and then to apply it to the structure of flowers.

IRRITABILITY OF THE FLORAL ORGANS.—Perhaps no parts of plants are more keenly sensitive to stimuli, or show a greater number and variety of results to excitement than flowers. A large proportion resemble plants which sleep, *i.e.* they exhibit movements according to the amount of light and heat which they receive. So various is this, that Linnæus was able to frame his floral clock. While many thus open their petals at definite periods and subsequently close them and die, as *Convolvulus*; yet a large number reopen them again when the due amount of light returns, like Daisies and other Composites, *Anagallis arvensis*, *Mesembryanthemum*, etc. Others, like *Silene nutans*, unroll their petals at night, but roll them up again by day.* Besides these spontaneous motions of the perianth, the stamens often exhibit movements, apparently without any external stimulus. Thus *Parnassia* and Saxifrages slowly move their stamens in suc-

* See Dr. Kerner's description of this flower. *Flowers and their Unbidden Guests*, p. 133.

cession, either towards the pistil as in the latter, or away from it as in the former. Other flowers, like *Cratægus, Rubus*, and *Alisma*, have them at first spreading away from, but afterwards bending over the pistil. These processes facilitate one or other kind of fertilisation, and are very common.

Slow movements of the filaments after the anthers have discharged their pollen, so as to place them out of the way of the pistil, are not at all uncommon in strongly protandrous flowers. *Echium** and *Teucrium Scorodonia*† will illustrate this well-known phenomenon. The "lemon-scented" or "oak-leaved" species of *Pelargonium* has small and very irregular flowers, somewhat papilionaceous in appearance, with the stamens declinate, lying on the anterior petal; the style lies beneath them, with the five stigmas quite undeveloped. After the anthers have shed their pollen, they fall off, and the filaments bend down outside the flower, while the stigmas now come to maturity and lie in the very place where the anthers lay before them.

Similar slow movements are very common in the styles and stigmas of plants. In the *Compositæ* and *Campanula, Lobelia, Gentiana,* etc., the style arms with their stigmatic papillæ curl backwards, and so secure self-fertilisation. In several of the *Scrophularineæ* and *Labiatæ*, the style gradually bends over, so that the stigma comes in contact with the pollen. This, however, may be partly due to prolonged growth. As examples, may be mentioned *Rhinanthus, Melampyrum, Galeopsis, Stachys sylvatica*, etc. Treviranus says the same thing occurs with *Gladiolus*, the style curving back towards the anthers.‡

* Cf. Figs. 20, *b* and *c*, p. 82: *b* shows the position of the stamens before pollination; *c*, after it.
† See Müller's *Fertilisation*, etc., p. 500, fig. 169.
‡ Ibid., p. 548.

160   THE STRUCTURE OF FLOWERS.

In addition to slow and seemingly spontaneous movements, to which all organs of a flower are liable, there are many rapid actions, brought about by the direct means of external stimuli applied to them. Thus *Indigofera* and *Genista* are two genera in which the claws of the petals are in a great state of tension when the flower is open, and the moment they are touched it explodes. The claws, from having been horizontal, curl downwards, and the staminal tube with the included pistil is jerked upwards. Thus Fig. 47, *a*, represents a flower of *G. tinctoria* just expanded. On passing a pencil point down the front of the standard, the wings and keel petals drop vertically, as seen in Fig. 47, *b*, looked at from the front. The staminal tube now lies against the standard. The keel, from its extreme tension, splits where it curls at the base, and becomes wrinkled in front, as seen in Fig. 47, *c*.

There is a plant of the order *Convolvulaceæ*, the corolla of which actually closes on receiving a mechanical touch. M. H. Dutrochet, after observing that the movements of *Mimosa pudica* and *Dionæa muscipula* are all in one direction only, as also of the stamens of *Cactus opuntia* and *Berberis*, adds: " Mais il est quelques cas où cette incurvation oscillatoire s'effectue dans plusieurs sens différents, tel est, par exemple, le phénomène que présente une plante du genre *Ypomœa*, observée aux Antilles par M. Turpin, plante encore

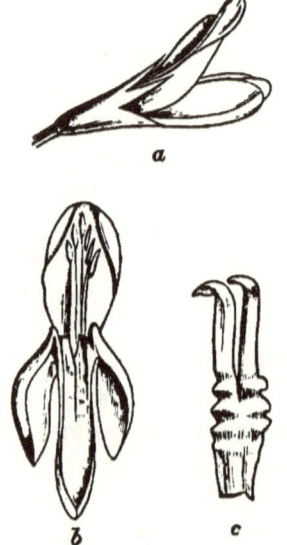

Fig. 47.—*Genista tinctoria:* a, before, b, after explosion; c, claws of keel.

inédite, qu'il désigne sous le nom d'*Ypomœa sensitiva*. Le tissu membraneux de la corolle campanulée, de cette plante est soutenu par des *filets* ou par des *nervures* qui, au moindre attouchement, se plissent ou s'*incurvent sinueusement*, de manière à entraîner le tissu membraneux de la corolle, laquelle, de cette manière, se ferme complètement ; elle ne tarde point à s'ouvrir de nouveau lorsque la cause qui avait déterminé sa plicature a cessé d'agir." * M. Dutrochet then observes that this phenomenon is in no way essentially different from the closing of the corolla of *Convolvulus*, to which *Ypomœa* is nearly allied, when it passes into the sleeping state, as does the calyx or perianth of the *Nyctagineæ*.

Lopezia coronata exhibits a curious and rapid movement

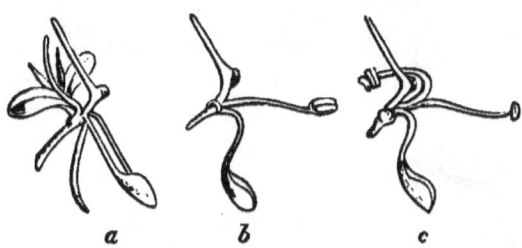

Fig. 48.—*Lopezia* (after Hildebrand). (For description, see text.)

in a staminode. Müller thus describes it : † "In each flower there is present one perfect stamen ; a second, standing immediately below, is reduced to a spathulate leaf, whose two halves fold upwards, and, in the first stage, projecting horizontally from the flower, inclose the anther of the perfect stamen (Fig. 48, *a*). The stalk of the spathulate leaf has an elastic tension downwards (*b*) ; the filament of the stamen an elastic tension upwards (*b*), so when an insect alights on the projecting spoon-shaped blade, as the only convenient

* *Recherches Anatomiques et Physiologiques sur la Structure Intime des Animaux et des Végétaux et sur leur Motilité*, 1824, p. 64.
† *Fertilisation*, etc., p. 265.

spot from which to reach two drops of honey that seem to rest upon a knee-shaped bend in the upper petals (*a*), the leaf springs downwards (*b*), and the stamen is set free and flies upwards, dusting the lower surface of the insect with pollen. When the stamen has thus served its purpose, it gradually curves upwards out of the flower (*c*), and the style which was hitherto undeveloped grows gradually out of the flower in a horizontal direction, so as to form another alighting place (*c*)."

Rapid movements in the stamens are not unknown. I described that of *Medicago* * many years ago, and now supply figures. Fig. 49, *a* represents the front view of a flower on expansion; *b*, the same after a bee has exploded it — the staminal column has now arisen, curled upwards, and abuts against the standard; *c* shows the curved position of the stamens, the corolla being removed. The stamens are inelastic, as they will not return to a horizonal position without breaking across, if pressed downwards.

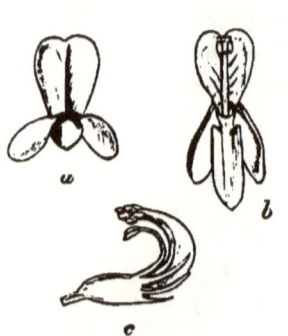

Fig. 49.—*Medicago sativa.* (For description, see text.)

Many other rapid movements of the filaments are too well known to need description, such as those of *Berberis, Helianthemum, Sparmannia, Centaurea,* and *Urtica;* while Orchids exhibit various movements in the caudicles of their pollinia.

Besides slow movements, the pistil often exhibits rapid ones on being touched, as are known to occur in *Stylidium, Canna, Maranta* and allied plants; while the flap-like stigmas of *Mimulus,*† and of several genera of orders allied to the *Scrophularineæ,* close together on being irritated mechanically.

* *Journ. Lin. Soc.,* vol. ix. p. 327.
† Mr. F. W. Oliver has lately investigated the mode of conduction

There is no need to describe a long series of movements, my object being simply to emphasize the fact that sensitiveness and irritability are pronounced phenomena in flowers, which point to a highly irritable condition of the protoplasm contained in the cells of all the floral members.* And, although we cannot now trace the progress of change in the floral organs under the mechanical and physiological impulses due to insect agency, the probability that these have been the actual influences to which the tissues have responded, and thence evolved the existing floral structures, will now, I trust, appear to the reader to be of a very high order.

of the irritation in the stigmas of *Martynia lutea* and *M. proboscidea*, and of *Mimulus luteus* and *M. cardinalis*. He believes it to be due to the continuity of the protoplasm from cell to cell. The tissue of the stigma consists of two lamellæ. The irritability is confined to several layers of prismatic cells on the inner side of the lamella, where the continuity of protoplasm was determined. (Quoted from *Journ. Roy. Micr. Soc.*, 1887, p. 781. Ber. Deutsch. Bot. Gesell., v. (1887), p. 162.)

Mr. Oliver has also lately contributed a valuable paper to the *Annals of Botany* (vol. i., p. 237, pl. xii., 1888), on "The Sensitive Labellum of *Masdevallia muscosa*." Continuity of the protoplasm occurs in the irritable "crest" on the labellum, which rapidly rises on being touched; the mechanism being closely comparable with that of the pulvinus of *Mimosa*. The author corroborates Mr. Gardiner's observation that a large amount of tannin occurs in the cells with which such irritability is concerned. References are also given to descriptions of other Orchids remarkable for having irritable perianths.

* For further information on the effects of light and heat upon the opening and closing of flowers, the reader is referred to Sachs' *Physiology of Plants*, chap. xxxvi., p. 641, where the author gives an account of Pfeffer's investigations. It is not clear, however, how temperature acts. A casual discovery may perhaps supply a hint. On forcing air into the flower-stalk of the white Water-lily, I found that the petals instantly spread open. May not, therefore, a rise of temperature cause the air within the tissues to expand, and so at least help to produce the same effect?

CHAPTER XVIII.

IRRITATION OF THE POLLEN-TUBE—THE ORIGIN OF CONDUCTING TISSUES.

The first effect produced by the action of the germination of the pollen-tube is the formation of the so-called conducting tissue or layers of specialized cells which nourish the tube in its downward growth. Like glandular nectaries, this tissue consists of small merismatic-like cells, highly charged with nutritive and saccharine substances. In some cases it is a metamorphosed condition of the epidermis alone, as

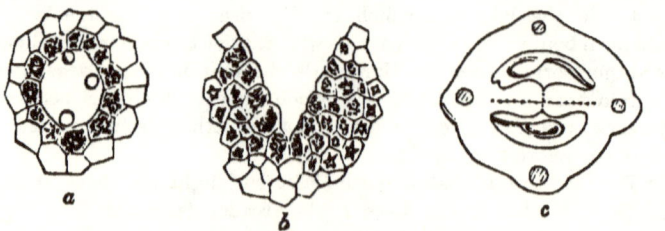

Fig. 50.—*a*, section of (epidermal) conducting tissue of *Fumaria*; *b*, that of *Rubus*; *c*, section of ovary of Crucifer (after Capes).

M. Capes has shown in his researches,[*] as in *Fumaria*. Fig. 50, *a*, represents a section of the stylar canal, the lining epidermis having its cells charged with such matters, while

[*] *Ann. des Sci. Nat.*, vii., 1878, p. 209.

three pollen-tubes are seen in section. Fig. 50, *b*, shows the formation of conducting tissue at the angle of the inflected carpellary edges of *Rubus*. The epidermal and subjacent cells form the conducting tissue in this case. The cells on the outskirts are charged with sphæraphids. Fig. 50, *c*, is a section of the ovary of a Crucifer. The replum or false dissipiment, as in the *Papaveraceæ*, forms the machinery for conducting the tubes. The dotted lines show the original lines of fusion. Now, if my theory be true, that no structure exists which has not been brought into existence through some foreign action having been brought to bear upon it—either directly from without, as insect agency, light, etc., or indirectly through nutrition within the plant,—then, the existence of this specialized tissue would never have arisen had it not been for the irritating action of the pollen-tubes. The analogous influence of the mycelium of a parasitic fungus here gives us the clue. As such causes hypertrophy to set in, and induces nutritive matters to accumulate upon which the fungus lives,—just as the irritation of the egg or pupa of a cynips or other insect causes a similar accumulation of richly nutritive substances to be made within the tissues of the gall upon which it feeds,—so the germinating power of the pollen-grain and the growth of the pollen-tube have actually brought about the formation of these highly nutritive conducting tissues of the style. The effect has then become hereditary, so that they are now in course of formation, at least, during the development of the flower in preparation for the ingress of the pollen-tubes.

The remarkably stimulating action of the pollen-tube had been observed more especially in Orchids. Hildebrand noticed that the influence of the pollen was twofold, in that it determined the growth of the ovary and the complete formation of the ovules *before* the process of fecundation had taken place.* M. Guignard has described the effects resulting from his experiments.† Thus, in the case of *Vanilla aromatica*, he found the development of the ovary was very rapid after pollinisation. At the time of flowering, the placentas have only the rudiments of the papillæ which will develop into ovules, and the conducting tissue formed by the epidermis and subjacent layers on either side of the placentary projections is still undifferentiated. In the intervals which separate the bands of conducting tissue, corresponding to the midribs of the carpels, there is no appreciable modifications before fecundation; but as soon as that has taken place, a layer of elongated papillæ, filled with a granular substance, arises. With regard to the development of ovules, M. Guignard remarks: "La pollinisation et la germination du pollen sont indispensables à leur formation. L'ovaire d'une fleur non pollinisée ne s'accroît pas et tombe quelques jours après l'épanouissement."

As soon, however, as the pollen-tubes are formed, the ovules begin to grow, until the twentieth day, when the primine thickens (much more than in other orchids) and finally gives to the matured ovule a globular form.

In the mean time the embryo-sac and sexual apparatus have been forming, and are completed (excepting the fusion of the two members of each tetrad, which does not take place to form the secondary embryo-sac nucleus) in little more than a month after pollinisation. Five weeks after that period, fecundation commences.

In following the progress of the pollen-tubes, it is not

* *Die Fruchtbildung der Orchideen, ein Beweis für doppelte Virkung des Pollen*, Bot. Zeit., 1863. *Bastardirungsversuche an Orchideen*, Bot. Zeit., 1865.

† *Ann. des Sci. Nat.*, 1886, tom. iv., p. 202; see also Maury, *Observations sur la Pollinisations des Orchidées*, comp. rend. de l'Acad. des Sci., 2 Août, 1886; and also Guignard, do., 19 Juillet, 1886.

till from the twelfth to the fifteenth day that some of them arrive at the base of the ovary. Before the sexual apparatus is complete, the extremities of the pollen-tubes separate from the mass of tubes overlying the conducting tissue, twist in various directions, scrambling over the placentary lobes and their ramifications, and so approach nearer and nearer to the ovular fucicles; but they only penetrate the micropyles, after the formation of the sexual apparatus. It is supposed by Strasburger that the synergidæ expel a liquid destined to guide the pollen-tube to the embryo-sac; others think their function is to aid in the solution of tissues for nourishment. In *Vanilla*, for example, the upper part of the embryo-sac is absorbed where occupied by the synergidæ, and is then covered by the elongated border of the primine. M. Guignard, however, adds:—"Il est possible qu'il soit attiré par un liquide expulsé par les synergides,* comme le pense M. Strasburger, ou bien aussi, comme je crois l'avoir constaté, par l'état spécial de la couche superficielle des membranes cellulaires du bord interne du tégument." †

On the action of the pollen-tubes M. Guignard writes as follows:—"Au contact des faisceaux polliniques, le tissu conducteur offre un contenu riche en sucre réducteur; l'amidon, dans le cas actuel, ne se trouve qu'au voisinage et du côté externe des faisceaux libéro-ligneux des parois ovariennes. Outre le pouvoir d'attaquer la substance amylacée et d'intervertir la saccharose, comme l'ont montré tout récemment M. Van Tieghem,‡ et M. Strasburger,§ les tubes polliniques peuvent aussi, à l'aide des ferments qu'ils contiennent,

* *Synergidæ* is better, being nearer *Sunergatai*.
† *L.c.*, p. 209.
‡ *Sur l'Inversion du Sucre de Canne par le Pollen*, Bull. Soc. Bot. de France, 1886.
§ *Ueber Fremdartige Bestaübung*, Pringsh. Jahrb., vol. xvii.

dissoudre la cellulose, ainsi que le prouvent les soudures avec fusion que j'ai observées plusieurs fois entre eux dans les cultures, où le phénomène est plus facile à voir. D'ailleurs, la pénétration directe des tubes polliniques dans les papilles du stigmate de plusieurs fleurs, après dissolution de la membrane cellulaire, est un fait du même ordre."

I quote this passage in full, that the reader may see how it completely corroborates my belief that the metamorphosis of the epidermis and subjacent layers to form the conducting tissue is entirely owing to the action of the tubes themselves, as well as the conversion of starch into saccharine, and therefore easily absorbable matters.

M. P. Maury has noticed very analogous facts in *Verbascum*, in that " at the period of pollination the ovules are still in a rudimentary condition, and altogether unfit for fertilisation. The nucellus is entirely occupied by the embryo-sac, in the protoplasmic contents of which there is as yet no differentiation of oosphere, synergidæ, or antipodals. It is only after the pollen-tube reaches the micropylar canal that these begin to be formed." *

This observation corroborates what I have said above, that not only is the pistil delayed in development in insect-crossing flowers, but that arrest of growth may affect all parts, and particularly the ovules; and I strongly suspect if more instances, of the *Gamopetalæ* especially, were examined it would be found to be the rule and not an exception. M. Maury's investigations also agree with M. Guignard's, in that the action of the pollen-tube is a stimulating one, and brings about developments which would not, and, indeed, cannot, otherwise take place.

In *Vanda tricolor pallens*, experimented upon by M.

* *Bull. Soc. Bot. Fr.*, viii. (1886), p. 529, quoted from notice in *Journ. Roy. Micr. Soc.*, 1887, p. 433.

Guignard, he noticed the not infrequent effect of a rapid change of colour in the perianth after pollination, although it did not fade for a week. The swelling began on the second day in the "gynostème," and progressed towards the ovary. From having been four centimetres long on the day of pollination, December 4th, 1885, by the 15th of April, 1886, it had grown to seven centimetres. The ovules, however, were not full grown, the embryo-sac having still its primitive nucleus; by the 15th of May, the ovules had attained their complete development. By the 1st of June, fecundation had taken place in nearly all the ovules. Hence about six months were required for the process.

In this species the spaces over the mid-ribs were covered with long hairs, corresponding to the papillæ in Vanilla. In both they appear to have grown after, and as a result of, pollination.*

In a flower of *Angræcum superbum* which became arrested the influence of the pollen-tube was remarkably illustrated. Three weeks after pollination an arrest of development followed in the ovary; it had sensibly increased in diameter in the upper part. On examining the ovarian cavity at the top, M. Guignard found only a small number of pollen tubes, relatively short in length.†

Another abnormal case was a *Vanilla*, in which, from some unknown cause, only two bundles of pollen-tubes were formed on either side of a placenta. Here the ovary grew on that side, causing a strong curvature. On the opposite side, the wall and the placentas with their ovules were atrophied.

* Max Wichura found that silky hairs were sometimes the sole result of his attempts to hybridize willows; and as analogous instances are the clothing the interior and exterior surfaces of galls with papillæ or hairs, an indirect result of the irritation set up by the pupæ (p. 138).

† A like interpretation may be given to Vegetable Marrows when they swell only at their distal end.

The exciting effect of the tubes is seen when Orchids are crossed which have no affinity, and are therefore incapable of fertilisation. Thus, the pollination of *Orchis mascula* by *Cypripedium parviflorum* even determined the formation of the sexual apparatus in the former. Similarly, when *Orchis* and *Listera*, as well as *Ophrys* and *Limodorum* were crossed, ovules reaching various degrees of development were produced, but none were impregnated.

Everything indicates (writes M. Guignard) that the development of the ovules is subordinated to that of the ovary. In exotic Orchids the thickness of this organ and its elongation are often very pronounced before the appearance of the ovules.*

Analogous results have been obtained by Max Wichura's experiments † in hybridizing Willows, who noticed the following degrees of failure indicating the various amounts of influence that the pollen-tube had over the sexual apparatus of the plants crossed : (1) the ovaries swell and ripen, but contain no seed ; (2) the ovaries are quite filled with silky hairs which clothe the umbilical cord end of the seed, but contain no embryo ; (3) seeds are present, but small and incapable of germination ; (4) seeds apparently perfect, but do not germinate ; (5) seeds germinate, but the seedlings are weak, and soon wither ; (6) seeds few but fertile and active ; (7) seeds numerous with only a few fertile ; (8) seeds numerous and fertile.

* Gærtner, in his *Mémoire sur les Organes Reproducteurs des Phanérogames*, devotes a special chapter to the enlargement of the ovary without previous pollination, with the result of a pseudo-fruit (*Versuche u. Beob. über die Befrucht. Organe der Vollk. Gewächse*, 1844).

† *Die Bastardbefruchtung in Pflanzenreich, erläutert an den Bastarden der Deiden*, Von Max Wichura. Mit zwei Tafeln. 4to., Breslau, 1865. Abstract by Rev. M. J. Berkeley, Journ. Roy. Hort. Soc., New Series, vol. i., p. 57.

Similar results occur in many cultivated plants without hybridizing; as appear in seedless Oranges; Grapes, such as "Sultanas;" Bananas, Cucumbers, etc.

Every other cause capable of acting in the same way will produce a like result, as in various instances of parasitism, when the cells become hypertrophied, as do those occupied by *Synchytrium*, or as in the roots invaded by *Plasmodiophora*. M. Guignard quotes an interesting case, which fully bears out the theory advanced in this book, of the results of the irritation of insects. He says, "À l'appui de cette manière du voir je citerai une observation intéressante que le hasard a fournie à M. Treub,* et dont ce savant a bien compris la signification réelle."

"Ayant rencontré des ovaires de *Liparis latifolia* qui présentaient un épaississement plus ou moins considérable, même dans les fleurs non épanouies, et où la pollinisation directe ou indirecte n'avait pu se faire, il trouva à l'intérieur des petites larves qui y avaient pénétré de très bonne heure. Ces larves ne paraissaient exercer aucune influence nuisible sur les cellules, et semblaient avoir la faculté de se mouvoir librement dans la cavité ovarienne, bien qu'on les trouvât au contact de la paroi ou des placentas. Elles se nourrissaient évidemment des sucs de l'organe envahi; à peine voyait-on une légère altération de quelques cellules avec lesquelles elles étaient en contact. Comparés à ceux des fleurs normales avant la pollinisation, ces ovaires habités par les larves offraient des placentas plus grands et plus digités, sur lesquels s'étaient développés finalement des ovules revêtus de leurs deux téguments formés comme sous l'influence de la pollinisation. Les dimensions des ovules ne différaient pas de ceux des graines mûres provenant d'ovaires pollinées, et non envahis par des larves.

* *Notes sur l'Embryon*, etc., Ann. du Jard. Bot. de Buit., iii., p. 121, pl. xix.

"Il était donc évident que les parasites avaient déterminé les mêmes effets que les tubes polliniques : l'accroisement des ovaires et des placentas et le développement des ovules."

The reader will here see the importance of this curious instance as bearing upon my general theory of growth in response to irritation; so that if ovaries, placentas, and ovules can be stimulated into growth and development, there is no *à priori* reason why other parts of flowers may not equally well grow in response to irritations set up by the insect visitors; as I have already shown to be the case in *Clerodendron** and in Mr. O'Brien's experiments.†

Perhaps it will not be amiss to notice here a very similar action of the suspensor in Orchids, described by M. Treub, which grows "backwards," escapes from the micropyle, and then ramifies in various ways, clasping and burrowing into the ovarian walls like a parasite in order to convey nutritive matters to the rudimentary pro-embryo.‡

Finally, M. Guignard remarks upon the degradations in the essential organs of Orchids as accounting for the well-known difficulty in raising seed from them : "Malgré le nombre immense des grains formées dans les conditions naturelles comme dans les serres, nombre qui paraît être d'ailleurs une signe de dégradation physiologique dans une famille où la différenciation morphologique des organes floraux est cependant si élevée, l'insuffisance de réserve alimentaire contenue dans leur embryon microscopique, en nécessitant des conditions spéciales pour le développement, suffit peut-être à expliquer les difficultés et les insuccès de la reproduction des orchidées par graines, et la parcimonie relative avec laquelle elles sont distribuées dans la nature."

* See p. 130. † See p. 114.
‡ *Notes sur l'Embryogénie de quelques Orchidées*, Verhandelingen der Koninklijke Akademie van Wetenschappen, 1879.

With regard to the difficulty of rearing Orchids, the reader may be referred to the Report on the Orchids Conference,* in which Mr. B. T. Lowne observes: "One of the difficulties in rearing seedling Orchids arises, I believe, from the fact that the pollen is only developed from the prolification of the mother cells, after the pollinia are placed on the stigma." He also found that, besides the pistil thus stimulating the pollen, "the stimulation due to the presence of the pollinia gives rise to the development of the capsule, even whilst the ovules remain unimpregnated." †

The significance of the above details lies in the fact that external influences, both mechanical and physiological, can bring about changes in the epidermal ‡ and sub-epidermal layers, with a determination of a flow of fluids of a specific character to those specialized tissues. As this is proved to be true for the conducting tissues, so do I infer it to be equally so for glands of various kinds.

* *Journ. of Roy. Hort. Soc.*, vol. vii.; see paper by Mr. H. J. Veitch, p. 22.

† *L.c.*, p. 48. "Degeneracy" will be discussed in Chaps. XXVI. and XXVII.

‡ M. Mer found that stomata were developed in the epidermis of galls on vine-leaves which normally had none. "Insolation" or exposure to light has a marked influence on the orm of the epidermal cells, and in increasing the number of stomata. The walls become straighter and thicker, and especially the cuticle. M. Mer believes the production of stomata to be the direct result of the accumulation of nutrient substances. *Comp. Rend.* xcv., 1882, p. 395. See also *Journ. Roy. Micr. Soc.*, 1882, p. 530, and 1883, p. 91. See above, p. 154. Another important paper on the same subject, fully corroborating these observations, has lately appeared, by M. L. Dufour, entitled, *Influence de la Lumière sur la Forme et la Structure des Feuilles*, Ann. des Sci. Nat., 7 sér., tom. 5 (1887), p. 311.

CHAPTER XIX.

COLOURS OF FLOWERS.

THE LAWS OF COLOUR.—M. de Candolle proposed to divide the colours of flowers into two series, the Xanthic and Cyanic, the former containing yellow-green, yellow, yellow-orange, orange and orange-red; the latter, blue-green, blue, blue-violet, violet, and violet-red; red being intermediate between the two series. It was thought that flowers were rigidly bound by these series, and never transgressed them, but that the tints of a species might vary through each. Thus the editor of the *Gardener's Chronicle*, replying to a correspondent on Feb. 2, 1842 (p. 97), remarks that "a blue Dahlia was not to be expected. On the other hand, the Hyacinth, being of the cyanic series, a yellow Hyacinth will not occur."

Yellow Hyacinths are, however, common enough now. Even in 1856, Dr. Lindley found it necessary to conclude a leading article on the subject with the words : "At all events the cyanic and xanthic speculations of philosophers must now be laid up in the limbo of pleasant dreams."

The many exceptions to this supposed rule met with between 1845 and 1856 elicited the above remark, and notably a species of *Delphinium*, viz. *D. Cardinale*, containing "golden yellow in the petals, which are as scarlet as a soldier's jacket everywhere else, one of the last of Messrs.

Veitch's fine Californian introductions. In this flower there is no sign of blue. Yet, if there is a genus more pre-eminently blue than any other cyanic race, it is surely *Delphinium.*"

It is true that some species have never yet transgressed their bounds, so that Dahlias still refuse to be blue now as in 1845; and we are still ignorant of the reason.

The effect of nutrition upon the colours of plants is well known, in that they vary much more in a garden soil than in the wild state; and differ in colouring according to the character and ingredients of the soil. Thus, as described by a writer in Hovey's *Magazine of Horticulture*,* striped Dahlias will be best kept clean by planting them in a poor soil, while rich soil invariably runs them. *E.g. D.* var. *striata formosissima*: No. 1 was planted in a poor gravelly soil, in an open situation; all the flowers but two were beautifully mottled. No. 2 was planted upon a rich, cool, sandy loam; not one-half of the flowers were mottled. No. 3 consisted of three plants, very highly enriched; every bloom but one was self-coloured. Similar effects follow on the variegated foliage of Pelargoniums, according as they are grown in a too rich soil or light one.†

"Alum is said to render the Hydrangea blue; and some saline substances, such as phosphate of iron and muriate of ammonia, appear to brighten the tint of red."‡ It often happens, however, that blue and pink corymbs occur on the same plant of *Hydrangea*. A cutting taken from a blue *Hydrangea* growing at Southampton, and transferred to Bedfont, changed to the usual colour on blooming there.§

Chloride of lime has been known to make a whole-coloured

* Quoted in the *Gard. Chron.*, 1842, p. 8.
† *Gard. Chron.*, 1876, p. 567.   ‡ *Ibid.*, 1843, p. 577.
§ *Ibid.*, 1886, vol. xxvi., p. 118.

Camellia become striped; while ammonia enhances the colours of Balsams.

Oxidization is believed to have great influence in changing the colours of plants, just as it affects certain juices when exposed to the air. Thus, if a leaf of the Socotrine Aloe be injured, the juice is at first violet in tint, but it soon turns to brown. If a potato be grated, the pulp rapidly browns in a similar way. Many fungi, especially noted for their poisonous properties, turn blue on injury, as species of *Boletus*. Moreover, they do not do so if exposed to nitrogen, hydrogen, or carbonic acid; hence it is presumably the oxygen which effects the change.

Some flowers change their colours from their first opening to a full expansion: such as *Cobæa*, from green to violet; several Boraginaceous plants, from red or even yellow to blue-purple. *Lycium barbarum*, the popularly called "Tea-plant," is a well-known instance. Others change during the day, as the "Changeable *Hibiscus*." This plant has flowers white in the morning, pink at noon, and bright red by sundown.* Similarly, a *Phlox* of a bright pink colour, "in the early morning, by five o'clock, has its colour of a lightish blue, which continues to alter as the sun advances, and by nine or ten o'clock becomes its proper colour; the clump which catches the sun's rays first changes first, while the other is still blue."

Though referring these and other well-known instances to oxidization, Dr. Lindley, from whose leading article the above remarks are partly taken, concludes with the observation, "In fact, we know very little about the cause of changes in colour, either in plants or animals." Perhaps it remains so still.

The intensity of the colours of many high Alpine flowers

* According to M. Ramon de la Sagra; quoted in *Gard. Chron.*, 1842, p. 555.

has often been noticed; and when plants growing near Paris were transferred to a much higher latitude, the flowers deepened in colour. This, however, is thought not to be due to a clearer atmosphere, but to the enhancement of the foliage, as M. Ch. Flahault showed that the leaves of plants of the same species are larger in proportion as the latitude is higher, the comparatively large dimensions being due to the duration of light of a relatively feeble intensity. Flowers being dependent upon leaves, great importance must be attached to the power of the latter to store up nutriment for them. Thus, in the case of Hyacinths both blue and red, M. Flahault found no difference in the colour of the flowers when grown in the light or in the dark, the colour being at the expense of the material stored up in the bulbs. Other experimenters have found that, while some flowers show no difference, others do; thus Askenazy found no difference in Tulips and Crocuses, though the leaves were etiolated in the dark. With Hyacinths, however, contrary to Flahault, he found that light exerted a two-fold influence, an acceleration of at least a fortnight in flowering and a much more intense and more diffused colour; those in darkness being only tinged where the uncoloured ones where darkest. *Pulmonaria officinalis* in darkness changed from red to blue, as usual; but in proportion as the buds were in a less advanced stage when placed in darkness, so were the colours fainter. His conclusion is that some flowers require light to develop their normal colours, while others are independent of it.* Mr. Sorby † agrees with Askenazy; and concludes that the arrest of normal development in darkness varies with the nature of the colouring matters, the effect being greater with the more easily decomposable substances. "Those substances which when dissolved out from the petals are the most easily

* *Bot. Zeit.*, Jan., April, 1876.   † *Nature*, April 13, 1876.

decolorised by exposure to light, are formed in relatively greater amount when the flowers are grown in the dark. This is easily explained if we assume that a higher vital power, depending on the presence of light, is necessary to overcome the more powerful chemical affinities of the less stable compounds."

The crossing of flowers is well known, and much practised by florists, to enhance the variety of tints. The interpretation is that crossing is a stimulating process, and provokes the petaline energy to a high degree.

From the preceding remarks it will be now gathered that colours, *per se*, are a result of nutrition; and that the prevalence of brighter colours in conspicuous flowers which are regularly visited by insects is due to the stimulating effects which they have produced, thereby causing more nutritive fluids to pour into the attractive organs.

Besides, however, this general result of brilliant colouring there are those peculiar and special displays of bright tints distributed in spots and streaks in certain and definite places only. These have been called "guides" and "path-finders," as they invariably lead to the nectaries. If the theory be true which I am endeavouring to maintain throughout this book, all these effects are simply the direct results of the insects themselves. The guides, like obstructing tangles of hair and nectaries, are always exactly where the irritation would be set up; and I take them to be one result of a more localized flow of nutriment to the positions in question.

Instead, therefore, of a flower having first painted a petal with a golden streak to invite the insect, and to show it the right way of entering, the first insect visitors themselves induced the flower to do it, and so benefited all future comers.

THE ORIGIN OF COLOURS.—Mr. Grant Allen has written

an interesting little book on *The Colours of Flowers*,* in which he expresses his belief that the first colour on departing from the primitive green was yellow. When we remember that the spore-cases and spores of *Lycopodium*, the anther-cells of *Cupressus*, and the whole anther-scale of Pinus and all the pollens of *Gymnosperms* are yellow,—again, when we come to Dicotyledons and find the prevailing tint of stamens is the same,—we gather probabilities in support of that view. That Nature next introduced reds, and only lately, so to say, succeeded in manufacturing blues, seems probable from the comparative rarity of the last colour. Moreover, when flowers individually change from one colour to another as they develop from the bud to maturity, it is always in that order—*i.e.* from reds to mauves or purples, as in *Echium*, *Pulmonaria*, etc., or even from yellows through reds to purples, as in *Myosotis versicolor*, so that we still seem to gather additional support to the theory.

If, however, we ask what has caused these changes, we are as yet in the dark. A few hints are attainable, and that is all. Yellows and reds seem to be due to substances allied to the oxidized products of chlorophyll in autumn leaves. Again, chlorophyll grains on turning yellow in fruits (*Lycium*) become angular, two or three pointed, and finally granular. In the same way the yellow granules of petals (*Cucurbita*) resemble "amyloplasts," or starch-forming corpuscles.†

The general conclusion one arrives at from various observations is that the original change from the ancestral green to, probably, yellow is correlated to the change of function; but why the first colour was yellow, and why it ever gave place to red or blue, is unknown.

Supposing the yellow-green colour to have spread to the adjacent parts which then attracted insects, as it does in

* In *Nature* series. † Sachs' *Veg. Phys.*, p. 320.

some *Euphorbias* and *Chrysosplenium*, for instance, then the visits of insects would bring the required stimulus to advance the colour to a pronounced yellow; and so petals, it may be conceived, came into existence.

PALE AND WHITE VARIETIES.—The paler tints or even a total absence of colour may seemingly occur as a variety of any plant. It is often a concomitant of habitual self-fertilisation in cases where the variety or species is a degradation from some conspicuous and brightly coloured insect-visited form. White-flowered individuals often appear as "sports" amongst seedlings; the immediate cause of which it would be difficult to assign, beyond the general one of the absence of those nutritive conditions which are requisite for colours, as occurs in *Gladioli*.*

White, however, is useful as a starting-point for florists' flowers where great variegation is required. Thus M. Vilmorin† says that "in ten examples of variegation which were produced under my own observation, the course was always the same. The original plant, with flowers whole-coloured, gave in the first instance a variety of flowers entirely white; afterwards, variegations were produced from this white variety on its returning towards the coloured type. ... This pure white variety usually gives in the first sowing a greater or less proportion of plants with flowers like those of the coloured type; but by careful selection through several generations the pure white type is in most cases completely fixed. ... It is only among the white varieties not completely fixed that the variegations make their appearance; at first they exhibit narrow pencillings, the coloured portion being only one-tenth, and sometimes only one-twentieth of the whole surface; but then in the following

* *Garden*, 1880, p. 327.
† *Flore des Ser. et des Jard. de l'Eur.*, (*Gard. Chron.*, 1852, p. 500).

COLOURS OF FLOWERS. 181

generation . . . the coloured portions begin to predominate. . . . I have never been able to observe a single instance of variegation coming directly from the coloured original. The contrary, however, takes place with regard to the dottings; these come directly from the coloured type." The variegated varieties the author had succeeded in fixing were *Gomphrena globosa; Antirrhinun majus, Convolvulus tricolor, Nemophila insignis, Portulaca grandiflora* and *Delphinium Ajacis*.

Other florists have found that by crossing whole-coloured flowers pure white seedlings may result.

Abutilons have an instructive and, in part, a somewhat similar history. No hybrids were raised from the old bronze-red and striped form, which was usually barren, until the white "Boule de Neige" was introduced. Mr. George crossed this with "Duke of Malakoff." The white one had itself previously thrown up every shade of dingy white; but whether by being spontaneously crossed or not, does not appear to be known. Some of the colours of the seedlings of this cross were pale and dark pink, pale orange, bright carmine, salmon, orange-red, etc.*

Somewhat analogous results were obtained by Mr. Veitch with Rhododendrons imported from Borneo. Thus a cross between the larger-flowered *R. Javanicum*, which is orange-coloured, with the smaller white narrow-lobed *R. Jasminiflorum*, gave rise to the rose-coloured "Princess Royal." A further cross of the last with the parent *R. Jasminiflorum* eliminated the red colour; the offspring, however, retained the form and size of the corolla of the "Princess Royal." It is called "Princess Alexandra."

In the above-mentioned the effect of the white has been *to separate the tints;* so that from the old Bronze-red *Abutilon Darwinii* we get yellows and reds of different

* *Gard. Chron.*, 1878, p. 792.

shades. Similarly the orange or buff-yellow *Rhododendron Javanicum* has been split up into various reds; the white having, so to say, eliminated the yellow.

The subsequent effect of crossing with regard to flowers is variety. With this fact florists and horticulturists are familiar: for as soon as crossed or hybrid seedlings are raised the varieties of colouring become infinite. It has been observed that the "spots" are more persistent than the base-colour of the flower. This fact agrees with the theory advanced that they have, whenever they occur as guides or path-finders, been determined by the insects and then become hereditary as much as the shape of the flower itself; and as that is maintained much more persistently than general colouring, so is that specialized colouring which has been equally due to insect agency.

With regard to the correlations which exist between colours and insect visitors, Müller especially has observed several. Thus beetles seem to affect yellows, *e.g. Thalictrum* and *Galium verum;* wasps and carrion insects, reddish-browns, such as of *Comarum, Epipactis,* etc., while the more intelligent bees, etc., delight in purples and blues; and it is thought that their selective agency has determined the survival of such special colours as they prefer. This has been probably the case, but we still want to know what is the immediate cause which induces one colour to change to another.

As high colouring or conspicuousness if the flower be white is due to insects, so pale colouring and inconspicuousness is due to their absence; but what the nature of the stimulus is we cannot tell. It enhances the assimilative powers; for the crossed plants, as Mr. Darwin abundantly proved, are usually larger plants. It usually infuses some of the characters, floral or foliar, of the male parent—but

COLOURS OF FLOWERS.

not always: several experimenters assert that, after every precaution, the offspring exactly resemble the maternal parent. But one rule florists always adopt in order to enhance the colouring is to use the pollen of the better-coloured plant, the maternal parent being usually the inferior one.

As an illustration of the relative effect, of crossing and self-fertilisation respectively on the production of colours, I quote the following passage from Mr. Darwin's work: *
"The flowers produced by self-fertilised plants of the fourth generation [of *Dianthus caryophyllus* or Carnation] were as uniform in tint as those of a wild species, being of a pale pink or rose-colour. Analogous cases [occurred] with *Mimulus* and *Ipomœa*. . . . On the other hand, the flowers of plants raised from a cross with the fresh stock which bore dark crimson flowers, varied extremely in colour. . . . The great majority had their petals longitudinally and variously striped with two colours."

Uniformity and paleness of tint are thus correlated with self-fertilisation; and since, whenever the latter process is persevered with, an increase of fertility follows, it is not surprising to find that such tints are usually accompanied by an increased power of seed-bearing. Thus, Mr. Darwin found that, "the proportional number of seeds per capsule produced by the plants [of *Dianthus*] of crossed origin, to those by the plants of self-fertilised origin, was as 100 : 125." Again, of *Antirrhinum majus*, the relative self-fertility of red and white varieties was as 9·8 : 20; of *Mimulus luteus* the same comparison gave the ratio of 100 : 147; while pale-coloured Pelargoniums are notoriously great seeders.†

---
* *Cross and Self Fertilisation*, etc., p. 139.
† For further illustrations, see my paper on *Self-fertilisation*, etc.

CHAPTER XX.

THE EMERGENCE OF THE FLORAL WHORLS.

THEORETICALLY, as already stated, a perfect flower should or might be composed of six whorls, if its parts be not spirally disposed,—the perianth, andrœcium, and gynœcium each consisting of two verticils. The very general rule for their emergence from the axis is centripetal. The subsequent rates of development of the several whorls may vary considerably, so that one part which emerged first, or at least very early, may be late or the last to arrive at maturity.

The calyx or outermost whorl of the perianth when present is nearly always the first to appear, and to grow rapidly to a relatively large size, and thus protects the more rudimentary parts within it; but if it ultimately remains rudimentary itself, or, it may be, is not entirely arrested, then it is the corolla which first emerges, the function of protecting the essential organs being relegated to it. Such is the case with the *Compositæ*, *Valerianeæ*, etc.

The corolla, with rare exception, emerges before the stamens, though it is very generally rapidly passed in development by the latter organs. In *Lopezia* and *Primula*, however, the stamens emerge first; and this has led some botanists[*] to regard the petals of the last-named plant as

---

[*] For references and literature on the structure of *Primulaceæ*, see Masters's paper, *On some Points in the Morphology of the Primulaceæ*, Trans. Lin. Soc., 2nd series, BOTANY, vol. i., p. 285.

THE EMERGENCE OF THE FLORAL WHORLS. 185

outgrowths from the stamens. My own observations tend to confirm those of Dr. Masters, that it is an exceptional fact, and not constant. It appeared to him " that in *Lysimachia Nummularia* the petals did really sometimes (but not always) precede the stamens in their development."

The stamens emerge before the pistil, and if there be two whorls to the andrœcium, it is the sepaline whorl which appears first; though the fully developed stamens sometimes assume a position, as already explained, within the petaline, as in *Geraniaceæ*. Like the corolla and staminal whorls, the carpellary appears all at once, and last of all.

With reference to the emergence of the individual parts of the whorls, it is an almost invariable rule that those of the outermost whorl of the perianth or calyx, if it consist of three or five parts, rise centripetally in succession according to the laws of phyllotaxis. Thus, if the calyx be pentamerous, its parts invariably emerge in quincuncial order, thus constituting a cycle of the $\frac{2}{5}$ type.. If it be trimerous, as in Monocotyledons, it is a cycle of the $\frac{1}{3}$ type. If, however, it be tetramerous, then the parts emerge in decussating pairs, as in *Tamarix tetrandra*, *Sparmannia*, *Philadelphus*, and the sepals in the *Cruciferæ*.* This clearly shows that a normally tetramerous calyx is the result of the combination of two pairs of leaves, corresponding to two nodes, the internode between the pairs being suppressed.

The parts of the inner whorl of the perianth or petals of the corolla, as also those of each staminal and carpellary whorl, almost invariably emerge simultaneously if the whorls be regular; though pronounced differences may occur in the case of irregular flowers. Similarly, when there is a strong spiral tendency, as in the *Ranunculaceæ*, members may arise

* The lateral sepals, though overlapped by the other pair, are the *first* to receive their vascular cords from the axis.

successively. If the stamens be very numerous they usually emerge in centripetal order, as in Buttercups; but they may form "centrifugal groups," as in *Hypericum*; the numerous stamens of *Cistus* and *Helianthemum*, as well as of *Cactus*, *Opuntia*, and *Mesembryanthemum*, and the *Loaseæ*, are also centrifugal in their development. Lastly, if the carpels form a whorl, they, too, emerge simultaneously; but if they be numerous and spirally arranged they emerge and develop in succession.

There are some additional points to be observed. The first is the method of change from tetramerous to pen-

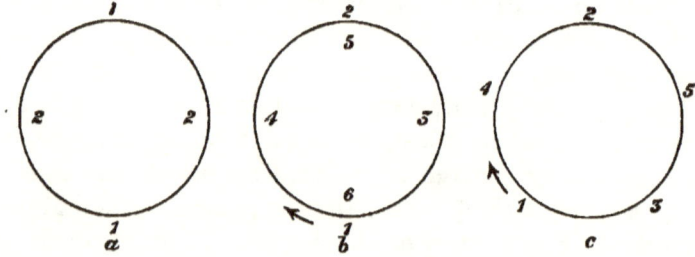

tamerous in the same plant. Thus in *Celastrus scandens*, if the flower be tetramerous, the sepals appear in pairs, the antero-posterior first, then the lateral pair afterwards. If the flower be pentamerous the sepals arise in succession quincuncially, the numbers 1 and 3 being anterior; numbers 4 and 5 are lateral, and number 2 posterior.

Now, by referring to the diagrams above, it will be seen that this order is in exact agreement with the usual method of passing from opposite to alternate arrangements in the foliage. The correct angular distance or divergence being acquired immediately in the case of the calyx, by shifting the position of the parts so that the divergence of 144° is obtained. In the case of foliage, this is only secured after several internodes (see p. 18).

THE EMERGENCE OF THE FLORAL WHORLS. 187

Exactly the same procedure occurs in *Sparmannia* and *Philadelphus*, which are tetramerous, as compared with *Tilia* and *Deutzia* respectively, which are pentamerous (see p. 18).

The next point to be noticed is the alteration in the order of emergence which takes place in irregular flowers. The rule seems to be that those parts of the flowers which assume a greater prominence in the mature state, or have some special function beyond the rest, emerge and develop before the others. Thus in *Leguminosæ* and *Labiatæ*, where there is a prominent "landing-place" for insects, the petals issue successively in an antero-posterior order. The carina of papilionaceous flowers composed of two petals appears first, then the alæ together, and finally the vexillum. In *Reseda*, the sepals, petals, and stamens issue in a postero-anterior manner; but while the sepals finally attain to much the same dimensions, the petals remain more or less atrophied as they emerge towards the anterior side. Then the stamens appear in the same order upon a cellular ring, which, later on, grows out into the unilateral disk between the petals and stamens.

In a few regular flowers the simultaneity is also wanting. thus in *Adoxa* the sepals of the tetramerous terminal flower emerge in pairs, and the four petals simultaneously; but in the lateral flowers the posterior sepals issue before the anterior; and of the five petals the posterior one emerges first, the two lateral secondly, and the two anterior ones last of all. These modifications are continued in the order of flowering. Thus the terminal flower expands first, and "all at once." Of the lower lateral flowers the two upper posterior sepals open out first, then the posterior stamens mature and shed their pollen. The anthers dehisce in succession from the lateral stamens, and lastly from the anterior ones. The lower sepals do not separate until after the upper stamens have shed their pollen.*

Though we are not in a position yet to account for all such deviations from general rules, yet I think in such cases as the *Leguminosæ* and *Labiatæ*, and probably *all* irregular flowers, that the rationale may with great probability be assumed to be the stimulus given from without to meet the extra strain which certain petals or stamens or both have to sustain while supporting the weight of an insect when visiting them. To meet this demand an extra supply of nutriment is sent to the parts which thus require it; and, in fact, I believe the final result has thus been actually brought about by the effort of the plant itself, so that it has developed parts in accordance with its requirements in a manner parallel with that which has obtained in the animal kingdom.

In the case of *Adoxa* I would regard the above-mentioned orders of development as a result of unequal distribution of nutriment in order of time. Thus the apical flower receives its nutriment first and develops first; then the other flowers which are placed laterally subsequently. And this order of supply has affected the parts of the latter flowers in the same way, so that they develop from above downwards, or in a postero-anterior manner. It may be compared to a three-flowered cyme, of which the central flower expands first, and the two lower ones afterwards.

A feature must here be noticed, though I do not think much stress need be laid upon it, which botanists have called "obdiplostemony." † If a flower have one whorl of stamens of the same number as the petals it is isostemonous; of two, diplostemonous; and if the stamens of the *outer* whorl be opposite or

---

* For a note on *Adoxa*, see my paper *On the Origin of Floral Æstivations*, Trans. Lin. Soc., 2nd series, BOTANY, vol. i., p. 194.

† Sachs' *Text-Book*, 2nd edition, p. 601.

THE EMERGENCE OF THE FLORAL WHORLS. 189

superposed to the petals, and therefore antipetalous, then the above term is used: for the rule is that the calycine whorl should be outermost and emerge first; then the petaline, which usually takes a position higher up on the axil; and, in at least most of the genera and orders where obdiplostemony has been noticed in the completely developed flower, it is simply due to the petaline whorl of filaments being, so to say, thrust outside the level of the calycine whorl by the protruding buttress-like bases of the carpels, as in *Geranium pratense.* This is still more the case in *Oxalis*, where, as in *Geranium*, the sepaline stamens become the taller set, the petaline the shorter; and the position of the former being more internal than usual, apparently in consequence of the appendages which grow on the outer side of the filaments.*

Again, the order of emergence may be the same as usual, namely the sepaline stamens first, then the petaline; but the position of the latter, instead of being within as is the rule, may be apparently on exactly the same plane as the sepaline, as in Heaths. Since, however, they do not emerge simultaneously, but one set is intercalated between the other, or even outside of it (Fig. 51), this order of appearance is, to my mind, a sufficient proof that they do not really belong to the calycine whorl.

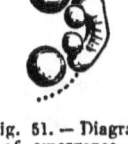

Fig. 51.—Diagram of emergence of petaline stamens of *Peganum*, outside the sepaline (after Payer).

There is no greater difficulty in understanding this, than in seeing that a compression of the internodes of opposite and verticillate leaves has taken place when double the usual number are present in a whorl. Thus privet has sometimes four leaves at one node, forming a quaternary whorl, and all on the same plane; and

* According to Frank, in *Oxalideæ* and *Geraniaceæ*, it is the antipetalous stamens which are developed first. See above, p. 150.

this will remind the reader that, since floral whorls are based upon phyllotaxis, ten stamens could not possibly form a cycle; and although the eight stamens of a Heath might do so, there is nothing in the leaf arrangement of that genus to suggest their being a whorl of the $\frac{3}{8}$ type.

Since the petaline cords are usually united to the staminal ones, the fact that the petaline stamens get sometimes, as it were, "dragged outwards," offers really no great difficulty; but is, so to say, a mere accident brought about by the adaptations of the flower to insect agency.

Indeed, to interpret these irregularities in the emergence, one must look to the final condition to see if there are any ultimate results in correlation with them. In *Oxalis* we get heterostylism with its corresponding different lengths of the filaments, and the necessary adjustments of the latter; since there are at least two sets in each flower, for insects to readily secure the pollen. In Heaths all the anthers are arranged in a ring round the style, pressing their cells against it, and so closely approximated, that when a bee dislocates one by pushing the lever-like auricle to one side, she dislocates the whole, and so receives a shower of pollen.

These final arrangements, therefore, are suggestive of the reason why the points of emergence of the stamens occur just where they do.

In the case of *Hypericum*, where the stamens emerge *centrifugally*, from a definite number of original papillæ, three or five as the case may be, the stigmas extend outwards; so that, if they have not been pollinated by insects, they can come in contact with the latest formed or the outermost anthers.

CHAPTER XXI.

THE DEVELOPMENT OF THE FLORAL WHORLS.

THE order in which the several whorls of flowers emerge from the axis is, as stated above, almost invariable; but the rates of development are very various, and important sexual and other differences follow as the results. For flowers with conspicuous corollas or other structures attractive to insects, the prevailing order of progression subsequent to emergence is first the calyx, secondly the stamens, and, if there be two series, the whorl superposed to the sepals grows first, afterwards the whorl superposed to the petals; then follows the pistil to a point approaching maturity, when the corolla, just before expansion, grows very rapidly to its full size; and finally the stigmas mature. The anthers have also grown long before the filaments, which at last elongate very rapidly. The usual result on maturity is various degrees of protandry, coupled with conspicuousness or attractiveness to insects. As a few of the examples I have examined may be mentioned *Ranunculus acris*, *Cardamine pratensis*, *Stellaria Holostea*, *Lychnis dioica* (male), *Malva moschata*, *Geranium* (larger flowered sp.), *Pelargonium*, *Tropæolum*, *Epilobium hirsutum*, *Œnothera biennis*, *Ipomœa*, *Veronica Chamædrys*, etc. In fact, this order of growth and development prevails generally with flowers having conspicuous corollas.

The interpretation appears to be as follows. In such

flowers as these, energy is especially directed into the development of the corolla and andrœcium; the former being large, and the latter supplied with much and often highly differentiated pollen. All this means the consumption of so much more nutriment; and, as the chief amount of floral energy is thus directed first into the andrœcium, then into the corolla—which often attains a far greater size than the other organs,—consequently these two whorls tend to draw a large amount of nourishment to themselves. In consequence of this, the pistil has, temporarily at least, to suffer; so that its growth is for a time delayed, and it does not mature as early as the stamens, which had, moreover, a considerable start in the race to maturity. Hence the result is that the stamens are often mature and even shed all their pollen long before the stigmas are prepared to receive it.

This, then, accounts for protandry being almost invariably the rule in the case of relatively conspicuous flowers.*

If flowers have two or more whorls or many series of stamens, as have many genera of *Caryophylleæ*, *Geraniaceæ*, *Ranunculaceæ*, and *Rosaceæ*, then the pistil may arrive at maturity *between* the periods of different series, or contemporaneously with some of them; so that, while the flower is protandrous with regard to the first stamens which mature, it is homogamous with others, and thus self-fertilisation can be readily secured if the flower fail to be crossed.

It may be here observed, though the fact will be dwelt upon again, that by far the greater majority of flowers, conspicuous or not, retain this provision for self-fertilisation; and that those flowers which normally cannot possibly fertilise themselves are in a very small minority.

* There are a few protogynous flowers, it is true, which are more or less conspicuous, but these exceptional cases have their own interpretations, which will be considered later on (see Chap. XXII.).

Nearly the same order of development as the above is maintained with some that have rather inconspicuous flowers in consequence of the corolla being small; but then it must be remembered that the other organs are proportionally small too, and, if they come at all, are visited by small insects. Such, for example, are *Malva crispa*, *Veronica serpyllifolia*, *V. agrestis*, etc. In these flowers, however, the pistil has a remarkably rapid growth as compared with the preceding cases. The cause is, that energy is now directed at once to that organ, instead of being so largely occupied by the stamens and corolla. The result is that the pistil matures more rapidly than in the previous cases, and sometimes even simultaneously with the stamens. The flower is therefore more nearly homogamous, and self-fertilisation can with them more easily ensue.

In many cases amongst inconspicuous flowers I could detect no appreciable difference at all in the rates of development of the essential organs. I would then describe the order as Calyx, Stamens + Pistil, Corolla. As examples are *Lepidium campestre*, *Sisymbrium Alliaria*, and *S. officinalis*, *Nasturtium officinale*, *Corrigiola littoralis*, *Œnothera bistorta*, etc. These are all, it will be noticed, very small-flowered plants. They are thus homogamous, and habitually self-fertilising.

The next order of development to be noticed is Calyx, Stamens, Corolla, Pistil. As far as my observations go, this order appears to be mainly confined to gamopetalous flowers, with a hypogynous corolla, as *Linaria minor*, *L. Cymbalaria*, *Veronica spicata*, *Primula*,* *Anchusa officinalis*, *Borago offici-*

* This order of development in Primrose has been observed by others, and apparently thought to be exceptional; so that the somewhat strange suggestion of the corolla being an outgrowth of the androecium was made by Pfeffer; but it by no means stands alone in this respect. See Sachs, *l.c.*, p. 609; *Jahrb. für Wissensch.*, Bot., vol. vii., p. 194.

nalis, Amsinckia angustifolia, Statice psilocladia and *Plantago Coronopus*, etc.

The remarkable delay in the progress of the development of the corolla during the emergence and first stages of development of the stamens is the peculiar feature. It sometimes allows the stamens to emerge first, as in *Primula*; or if they be nearly simultaneous, then the corolla may be suddenly checked, as in *Veronica*. But many differences occur; thus they emerge and grow up together in *Samolus*, while in *Anchusa officinalis* the corolla rapidly exceeds both stamens and pistil. In the case of *Amsinckia* the corolla and stamens appear to emerge almost together, and then follows the pistil, which the former quickly exceed in height. Then the pistil regains the height of the stamens, and they ultimately mature together. A similar procedure obtains with *Plantago Coronopus*: though the petals emerge first, the anthers quickly outstrip them, and the corolla grows considerably more than the pistil, which is consequently delayed; but when they are nearly developed and the corolla becomes scarious, then the style elongates with great rapidity, and the stigmas mature first, so that the flower is ultimately protogynous. Exactly the same course is followed by the floral whorls of *Statice psilocladia*.

The next order of development is Calyx (if present), Corolla, Stamens, Pistil; or even Corolla, Calyx, Stamens, Pistil. The cause of the corolla developing so soon is the arrest of the calyx, as in *Umbelliferæ, Valerianeæ*, and *Compositæ*. The corolla now has to act as a protecting organ, and always keeps in advance of the essential organs. Indeed, in the orders with epigynous and gamopetalous corollas, in which the calyx is usually obsolete or nearly so, the corolla actually emerges before it.

The last order of development to be mentioned in the case

of flowers possessing a corolla is Calyx, Pistil, Stamens, Corolla. As examples, I find the following illustrate this condition: *Ranunculus sceleratus, Cardamine hirsuta, Cerastium glomeratum, Arenaria trinerva, Sagina procumbens, Spergularia marina, Polycarpon tetraphyllum, Trifolium minus, Epilobium montanum, Gaura parviflora,* etc. This appears to be the most general condition for very small and inconspicuous flowers which are regularly self-fertilised. The interpretation is the exact converse of the order of development first described; namely, of the whorls of conspicuous flowers. All the above are inconspicuous, many being rarely if ever visited by insects; and as the corolla is minute, no nourishment is required for the petals, the stamens are often reduced in number and the quantity of pollen diminished. The pistil at once proceeds to grow, and the result is, if not homogany, protogyny.

It must be now borne in mind that the above differences in the order of growth and development must not be regarded as at all absolute or invariable, but only general rules as to what takes place; for the rates of growth of the respective whorls may vary in the same species according to external circumstances; so that a plant may be protandrous at one time or place, homogamous or even protogynous elsewhere or in another season, as the case may be. Indeed, Müller frequently calls attention to this fact, to which I shall have occasion to return.

EMERGENCE AND DEVELOPMENT OF THE OVULES.—If the ovules be tolerably numerous, the order in which they appear is not constant. It may be either from above downwards or from below upwards on the placenta. Thus, as Payer has shown by his drawings, in *Viola, Reseda, Cistus, Tetrapoma, Fumaria, Linum, Ruta, Melianthus, Staphylea, Spiræa,* and *Opuntia* the order is basifugal, or from below upwards. On the other hand, in *Macleya, Dicentra, Epimedium, Bartonia, Impatiens, Lythrum, Dracophyllum, Malachium, Cerastium, Primula*, and *Samolus* the order is basipetal, or from above downwards.

When the row of ovules is very numerous, it is the rule that the point where they first begin to emerge is midway, and the development takes place both upwards and downwards simultaneously. It is thus with *Helleborus* and allied genera with follicles, *Capparis, Epilobium, Trifolium, Cajophora, Lathyrus, Citrus, Passiflora*, and the Monocotyledonous orders, *Iridaceæ* and *Amaryllidaceæ*. *Lythrum* and *Opuntia*, however, both of which have considerable rows of ovules, develop them, as stated above, in a basipetal and basifugal manner respectively.

On examining Payer's numerous figures, I find that when the order of development is from below upwards, the ovules have their micropyles upward; when they develop from above downwards, the micropyles grow downwards. In either case, occasionally the middle ones may be somewhat horizontal, if they are somewhat numerous, as in *Bartonia, Spiræa*, and *Staphylea*. When they are very numerous and develop both ways from a point midway, then the ovules may either turn upwards or downwards; the majority being downwards in the proportion of nine to five.

As a theoretical interpretation to account for the general fact of the central ovules developing first when there are long rows of them, it may be due to the carpel being comparable to a lanceolate leaf, where the longest and therefore the most vigorous nerve-branch of the pinnate nerves is in the middle. If the rows of ovules emerge from below upwards, the carpel may be comparable to a more primitive type, as of monocotyledons with a palmate foliage. Thus the only exceptions I can find in Payer's figures of Monocotyledons are the *Gladiolus* and *Alstrœmeria*, where they are very numerous and follow the rule of commencing to emerge in the middle, and then proceed upwards and downwards. Though parietal placentas seem generally to have their ovules developed from below upwards, yet, as seen above, it is not uncommon with an axile placentation. If any interpretation be sought, I should feel inclined to associate it somewhat with a more primitive state of things, since a parietal placentation presents a more rudimentary character than an axile. But *why* they are developed thus, sometimes upwards, sometimes downwards, or both ways at once, is at present as inexplicable as the fact that leaves develop both basipetally and basifugally, either in their entirety, or as to their lobes and notches, which may be formed on either plan. Perhaps there may prove to be a common cause for both.

CHAPTER XXII.

HETEROGAMY* AND AUTOGAMY.

PROTANDRY, PROTOGYNY, HOMOGAMY, AND CLEISTOGAMY.—These conditions prevail in nature in varying degrees of freqnency. The first is common to all conspicuous flowers habitually visited by insects, and is accompanied by heterogamy. The fact that anthers mature their pollen before the stigmas of the same flower are ready to receive it, is due to the extra stimulus given to the androecium, which mostly effects simultaneously the enhancement of the corolla or perianth which attracts the insects (see p. 191). Like everything else in nature, it is very far from being absolute, and any flower may be protandrous at one time or place, while it may at another mature the essential organs together, and then it becomes homogamous, or it may be even protogynous.

These latter conditions prevail in less conspicuous flowers and all those which are fluctuating between a condition

---

* *Heterogamy*, i.e. union by intercrossing different flowers.
Autogamy, i.e. union by self-fertilising one and the same flower.
Protandry, i.e. stamens maturing the pollen before the stigmas of one and the same flower are ready to receive it.
Protogyny, i.e. pistil maturing the stigmas before the pollen of one and the same flower is shed.
Homogamy, i.e. pollen and stigmas of one and the same flower maturing simultaneously.
Cleistogamy, i.e. autogamous within an unopened perianth.

requiring insect agency and self-fertilisation or autogamy; as well as in the majority of flowers which are too inconspicuous to invite insects at all, or which never expand. The series of such flowers terminates in perfect and perpetual cleistogamy.

The first condition, or Protandry, does not now require special discussion or illustration; as it is the prevailing one in most conspicuous flowers: though it must be distinctly borne in mind that the exceptions are rare in which a flower cannot fertilise itself at some period or other before it fades; even though a large order, as *Orchideæ*, may furnish many examples.

Protogyny may arise from several causes. Müller has mentioned about twenty species of plants irrespective of the Grasses which are more or less decidedly protogynous; and what one notices is that many are Alpine species of genera which have other species dispersed elsewhere that are homogamous or protandrous. Thus *Anemone alpina* is protogynous, but *A. Narcissifolia* is protandrous. *Ranunculus montanus, R. parnassifolius, R. pyrenœus* are all protogynous. These may be compared with the smaller-flowered forms of *R. aquatilis* which are homogamous; but *R. flammula, R. acris, R. repens* and *R. bulbosus* are protandrous with the outermost stamens only. Thus, this genus supplies a progressive series. Other protogynous and mountain species are *Dryas octopetala*, species of Saxifrage, as *S. androsacea* and *S. muscoides*, and *S. Seguieri*: but Müller found *S. oppositifolia* and *S. tridactylites* to be sometimes feebly protandrous, at others protogynous. On the other hand, *S. rotundifolia, S. aizoides*, etc. are protandrous. *Loiseleuria procumbens, Trientalis Europœa, Bartsia alpina, Hutchinsia alpina*, and *Thalictrum alpinum* are all protogynous.

Secondly, a group of plants, the flowers of which have

the habit of blossoming early, as in the spring or the beginning of the summer, are protogynous; such are species of Hellebore, *Prunus*, and *Cratægus*, as well as the Horse-chestnut and *Mandragora vernalis*.

Some species are characterized by the habit of living in shady places, as *Geum urbanum* and *G. rivale*, *Chrysosplenium oppositifolium*, *Gagea lutea*, *Paris quadrifolia*.

Lastly, others have minute flowers, as *Geranium pusillum*, *Veronica serpyllifolia*, *Tofieldia*, and many other species, some of which I have mentioned when treating of the emergence and development of the floral whorls, where I have explained the cause. *

Wind-fertilised or anemophilous flowers are for the most part protogynous; for these flowers have been accompanied by strong degeneracy of the corolla and pollen, while all traces of nectariferous structures are almost invariably and entirely suppressed.† Hence *Thalictrum minus, Poterium. Sanguisorba, Plantago sp., Callitriche, Myriophyllum, Artemisia, Chenopodium, Amentiferæ, Juncaceæ,* and *Gramineæ* are all more or less characterized by being protogynous while they are anemophilous as well.

If we are not in a position to trace the actual causes of protogyny in every instance, we can at least see several influences which can bring it about. Temperature will be seen hereafter to be a most potent one; for a relatively lower temperature very frequently checks the energy of the corolla and stamens, without having any necessarily corresponding effect on the pistil, and several compensating processes then come into play; so, conversely, the pistil now gains the ascendancy and can mature first. This, therefore, will

---

* See Chaps. XX. and XXI.

† Intercrossing by insects may be recovered in anemophilous flowers; when honey may be again secreted, as in *Salix capræa* and *Sanguisorba officinalis*; see *Fertilisation*, etc., p. 236, fig. 77.

account for some mountain species, as well as those blossoming early or in shady places, being protogynous.

It must not be regarded as universally true. If flowers so situated or circumstanced be abundantly visited by insects, they will respond to their influence; and the consequence is, that many Alpine plants are even strongly protandrous, as well as spring-flowering plants and some which grow in shady places, as *Sanicula Europæa, Odontites serotina*, etc. It is when we compare the protogynous species with others of the same genus, that the influences of a lower temperature, shade, etc., more especially suggest themselves as true causes of protogyny in some species, while others may be homogamous or protandrous.

Many plants normally provided with conspicuous flowers, but accidentally growing in shady places, may often be found having them half opened or as quite closed buds, and yet fully fertile. The same occurs late in the season, when the flowering period is drawing to a close. Such flowers represent the preliminary stages leading to a permanently homogamous or protogynous condition, as the case may be, which are mostly autogamous as well.

Whatever may be the direct cause, and there may be others besides those I have mentioned, protogyny is easily brought about temporarily in individuals, or it may become hereditary and a permanent feature.

It need now hardly be added that, before protogyny is reached and emphasized, all degrees of passage can be met with from strong to weak protandry; then homogamy is acquired: and, after passing through oscillating conditions, permanent protogyny can be finally the result.

Many individual plants vary in this respect, being sometimes or in some places in one condition, and at other times and in other places in another condition. As nothing is absolute in nature, so in this case, plants respond to the influences brought to bear upon them, and each individual may vary accordingly, but if the influence be permanent, then the variation becomes hereditary, and one or other character is fixed, and may be regarded as specific or generic as the case may be. Should the environment change again, what may have been constant for generations will be once more broken up, and instability ensues.

Müller records several cases of such oscillations, as in *Pulsatilla vernalis*, *Dryas octopetala*, *Ribes petræum*, *Gentiana campestris*, *Veronica serpyllifolia*, *V. spicata*, Walnut, Hazel, etc. These vary from protandry through homogamy to protogyny. He also mentions species which have not yet arrived at complete protogyny, such as *Sibbaldia procumbens* and *Ranunculus alpestris*, mountain species which are homogamous; while *R. glacialis* is sometimes even slightly protandrous. *Papaver alpinum*, *Arabis alpina*, and *Biscutella lævigata* are also described as homogamous.

As the transitions from a conspicuous, protandrous, and entomophilous or insect-fertilised flower to a homogamous and autogamous or self-fertilised one, as well as to anemophily, are the effects of degeneracy, they will be considered more fully when that peculiar condition of floral structure comes to be discussed.*

* See Chaps. XXVI. and XXVII.

CHAPTER XXIII.

HETEROSTYLISM.*

DIMORPHIC FLOWERS.—A large portion of Mr. Darwin's work on the "Forms of Flowers" deals with the varieties and phenomena of heterostylism, which is specially characteristic of the *Primulaceæ*, and *Rubiaceæ*, though several instances exist in other orders as well. He and Mr. J. Scott were mainly interested in showing that "illegitimate" or homomorphic unions were less prolific than "legitimate" or heteromorphic; and inferentially took occasion to describe the differential sexual characters of the forms of the same species. With regard to this latter fact, when Mr. Darwin experimented with wild Cowslips, he first thought that they were tending towards a diœcious condition, and that the long-styled plants were more feminine in nature, and would produce more seed : conversely, that the short-styled plants were more masculine.

Contrary to his anticipation, of plants marked growing in his garden, in an open field, and in a shady wood, the short-styled forms gave most seed, the weight of seed being in the proportion of 41 to 34; that is, the short-styled produced more seed than the long-styled in the proportion of nearly 4 to 3. Similarly when a number of wild plants were

* *Heterostyled*, i.e. plants with stamens and styles of different but corresponding lengths on separate plants.
Homostyled, i.e. when stamens and styles are of the same length.
Homo-, di-, tri-, poly-, and *hetero-morphic, i.e.* flowers of the *same, two, three, many,* and *different* forms, respectively.

transferred to his garden, the result was as 430 to 332, the weight of seed being therefore nearly 4 to 3. Lastly, of plants covered by a net, six short-styled plants bore about 50 seeds, while 18 long-styled plants bore none at all.

From these results, Mr. Darwin wrote, " we may safely conclude that the short-styled form is more productive than the long-styled form. . . . Consequently my anticipation that the [long-styled form] would prove to be more feminine in nature, is exactly the reverse of the truth."* We shall see, however, that his surmise was probably, to some extent, right, nevertheless.

Mr. Darwin and Mr. Scott have recorded a great number of experiments in crossing heterostyled plants, and the following tables, constructed from details given by those authors, show to what extent the plants named were benefited by crossing either way.

LEGITIMATE OR HETEROMORPHIC UNIONS.

	Long-styled.		Short-styled.	Difference.
Primula veris (Wt. of seeds of 100 capsules)	†62	is to	44	18
P. elatior (Av. No. of seeds per capsule)	46·5	,,	47·7	1·2
P. vulgaris ,,	†66·9	,,	65	1·9
,, var. alba [Scott] ,,	19	,,	21	2
P. Sinensis ,, ,,	50	,,	§64	14
,, [Hildebrand] ,,	41	,,	44	3
P. Auricula [Scott] ,,	73	,,	§98	25
P. Sikkimensis ,, ,,	35	,,	§42	7
P. cortusoides ,, ,,	51	,,	§61	10
P. involucrata ,, ,,	66	,,	69	3
P. farinosa ,, ,,	52	,,	56	4
Hottonia pal. [Müller] ,,	†91·4	,,	66·2	25·2
Pulmonaria off. [Hild.] ,,	1·3	,,	1·57	0·27
Mitchella repens ,,	†4·6	,,	4·1	0·5
Linum grandiflorum ,,	5·6	,,	4·3	1·3
L. perenne ,,	7	,,	8	1
L. flavum (3 flowers produced capsules)	1	,,	3	2

* *Forms*, etc., p. 20.

The first observation is that in twelve cases the short-styled are in excess of the long-styled, and in four cases (†) this is reversed. Hence Mr. Darwin's conclusion is not absolute; and it is a somewhat remarkable fact that *Primula veris* (the Cowslip) is the identical species from which he deduced the conclusion that the short-styled was the more feminine of the two forms. The conclusion now arrived at from this species would be, that when it is *left to itself* the short-styled form sets most seed; but when *artificially crossed* it is the long-styled form which bears best. The cause of the former result is that some pollen in the short-styled form can fall upon the stigma and so secure self-fertilisation, which is impossible in the latter case. The same results occurred with Mr. Scott.*

Hence Mr. Darwin's first conclusion, that the short-styled was the more feminine, was drawn from a wrong premise; as it was not a question of *sex* so much as of *union*. When the results of self-fertilisation are compared, as given in the table on next page, it appears that the long-styled form of the Cowslip is the more feminine of the two, in the proportion of 42 to 30.

Of that table, three cases of *Primula sp.* (†) only show the short-styled bearing more seed than the long-styled when *illegitimately* fertilised; viz., with Mr. Scott, *P. vulgaris*, var. *alba*, and *P. Auricula* (*i.e.* forms more or less modified by cultivation); and with Hildebrand, *P. Sinensis*, when crossed

* *Journ. Linn. Soc. Bot.*, vol. viii., 1864. This case may be taken to illustrate one of the disadvantages often accruing through great differentiation and adaptation to insect visitors. Though it appears proved that legitimate crossing sets most seed when carefully and artificially effected; yet, when the process is left to the capricious visits of insects, Mr. Darwin's experiments show how nature fails to derive the full benefit of intercrossing; so that the Cowslip has to be contented with the results of the *illegitimate union of the least fertile* of the two forms.

by distinct plants. The difference, however, being only two in each case, is practically inappreciable.

Of the other genera, *Linum* shows a slight inclination in favour of short-styled; but as this genus is exceedingly barren when illegitimately fertilised, the results here given of that plant are insufficient for deducing conclusions; at all events, these tables show that the *long-styled form is certainly more prolific when illegitimately fertilised*, than the short-styled form when similarly treated.*

ILLEGITIMATE OR HOMOMORPHIC UNIONS.

			Long-styled.	Short-styled.	Differ-ence.
Primula veris	(Wt. of seeds of 100 capsules)		42	30	12
P. elatior	(Av. No. of seeds per capsule)		27·7	12·1	15·6
P. vulgaris		,,	52·2	18·8 [1]	3·4
,, var. alba [Scott]		,,	11	†13	2
P. Sinensis	,,	,,	35	25	10
,, [Hild.] (plants distinct)		,,	18	†20	2
,, ,, (same flower)		,,	17	8	9
P. Auricula	[Scott]	,,	12	†14	2
P. Sikkimensis	,,	,,	14	8	6
P. cortusoides	,,	,,	41	38	3
P. involucrata	,,	,,	38	28	10
P. farinosa	,,	,,	30	19	11
Hottonia palustris [Müller]					
(plants distinct)		,,	77·5	18·7	58·8
,, ,, (same flower)		,,	15·7	6·5	9·2
Pulmonaria off. [Hild.]		,,	0	0	0
Mitchella repens		,,	2·2	2	0·2
Linum grandiflorum		,,	2·5	†4·2	1·7
L. perenne		,,	0	†3	3

[1] "Too low"?

Referring to the column of Differences in the first table, it will be noticed that *two* of the four marked (†) of the long-styled are considerable, namely, *P. veris* and *Hottonia*; but the

* Mr. Darwin noticed that this was the case with the genus *Primula* (*l.c.*, p. 48).

HETEROSTYLISM. 207

other two are practically inappreciable. On the other hand, considering every difference under 5 as inappreciable, there are *four* cases (§) of the short-styled in which it is considerable; and of these it was only 3 in the case of *P. Sinensis* with Hildebrand; consequently one cannot confidently say that the short-styled is more feminine than the long-styled—at least, to any well-marked extent.

With the corresponding column in the second table, one notices *nine* cases where the difference is great; while in all of those marked (†) it is inappreciable. Hence the conclusion is much more pronounced in favour of the greater fertility of the long-styled forms when illegitimately crossed.

Müller accounts for "the greater productiveness of illegitimate crossings in the case of the long-styled form of *Hottonia* than in short-styled flowers, to the fact that the former kind of illegitimate crossings occur frequently in nature; as these flowers are visited by pollen-seeking flies which have no need to thrust their heads into the flower of the short-styled form," which is, therefore, presumably neglected.*

The table I have here drawn up shows that the greater fertility of the long-styled form when illegitimately fertilised, is a general feature of heterostyled plants, and not peculiar to *Hottonia palustris;* hence we must look to a more general cause.

As another hypothesis, it may perhaps be suggested that, as the homomorphic condition of *short stamens* with a *short style* seems to have been the primitive form, then in the

* If Müller's idea be true, *Hottonia* furnishes another instance of the disadvantage of great differentiations, and is only one degree better off than the Cowslip. In either case, one is inclined to ask what has become of the proper insects (whatever they may be) required for the perfect intercrossing of these flowers.

long-styled form the stamens are unchanged, while the pistil has elongated; whereas, in the short-styled form, with now elevated stamens, these and their pollen have presumably become differentiated, while the pistil has remained unchanged. Now the above result appears to indicate the fact that the long-styled *pistil* has not become physiologically differentiated to so great an extent as the *pollen* of the long-stamened form. The result is that it can be fertilised by the unchanged pollen of the same form more easily than the short-styled primitive form of pistil by the more highly differentiated pollen. This is not stated as a proved fact, and must be only regarded as a hypothetical suggestion. The extreme limits of differentiation are reached when the flower is heterostyled in form but diœcious in function. Thus *Ægiphila obdurata* seemed to Mr. Darwin to be in a diœcious condition, but derived from heterostylism, in which the long-styled was apparently female, and the short-styled male.

The species which shows the most marked difference between the produce of the legitimate fertilisation of the two forms is *P. Auricula* (or cultivated vars. of *Auricula*). It had been asserted by Prof. Treviranus that the long-styled unions were absolutely barren.* Mr. Scott shows that this idea arose from the fact that the plant in question had not been crossed. His experiments prove that the short-styled is the most fertile, whether legitimately or illegitimately crossed, though in the latter the difference is slighter: in the former the ratio being 8 to 6; and in the latter, 7 to 6.

Homostyled forms of *P. Auricula* are not uncommon. Mr. Scott found that 9 capsules gave 272 seeds, or an average of 30 seeds per capsule. Comparing this with the following results, its extreme fertility becomes apparent:—

* Scott, *l.c.*, p. 90.

Short-styled × homostyled gave 8 seeds per capsule.
Short-styled × short-styled ,, 14 ,, . ,, ,,
Long-styled × homostyled ,, 5 ,, ,, ,,
Long-styled × long-styled ,, 12 ,, ,, ,,

The pollen of the homostyled resembled that of the long-styled in appearance, though the stamens were situated high up as in the usual short-styled form. This seems to corroborate what was said above; for we have here also a long pistil fairly fertile with undifferentiated pollen.

Another species of *Primula* which often bears homomorphic flowers is *P. Sinensis*. Mr. Darwin's attention was first directed to it by observing a long-styled plant—descended from a self-fertilised long-styled parent—with the stamens low down but with the pistil of the short-styled form, though the length of the style varied in different flowers on the same umbel. He fertilised eight flowers with their own pollen, obtaining five capsules with an average of forty-three seeds. The examination of the pollen of two equal-styled plants showed a vast number of small shrivelled grains. In the case of two *white-flowered* plants, in which the pistil was neither properly long-styled nor short-styled, the size of the grains was in the proportion of 100 to 88; whereas, between perfectly characterized long and short-styled plants it would have been 100 to 57.

Of the first-mentioned homomorphic plants, four spontaneously yielded 180 capsules, with an average of 54·8 seeds, one containing 72; a result higher than could be expected of either form if self-fertilised. The next generation proved to be *all* equal-styled, *i.e.* the grandchildren of the four original plants. One of these bore an average of 68 seeds per capsule, with a maximum of 82 and a minimum of 40. Thirteen capsules, spontaneously self-fertilised, yielded an average of 53·2 seeds, " with the astonishing maximum, in one, of 97 seeds. In no legitimate union has so high an average of 68 seeds been observed by me, or nearly so high a maximum as 82 and 97." *

I give these results of homostyled Auriculas and Chinese Primroses as illustrating the principle so abundantly proved amongst other plants—that as soon as they begin to retrace their steps from a prevailing differentiated condition self-fertilisation is rapidly resumed, and there follows a resumption of a vastly increased rate of seed-making. They prove, too, that however apparently stable these highly differentiated states may normally be, various conditions of environment can readily break them down; thus, with cultivated plants, usually so much stimulated, starvation is a potent cause.† *Linum perenne*, as the above table shows, is particularly barren when illegitimately fertilised, but a single branch on a plant has been known to become homomorphic, and then to set seed abundantly; this occurred with Mr. Meehan. Warming found *Menyanthes trifoliata* to have become completely homostyled in Greenland.

TRIMORPHIC FLOWERS.—As a type of heterostylism where a species adopts three forms, *L. Salicaria* may be taken. Briefly summarizing Mr. Darwin's elaborate experiments on

* *Forms of Flowers*, pp. 218–221.

† It is not only true with heterostyled plants, but the rule applies generally to highly cultivated flowers, that degeneracy from a floral point of view is correlated with enhanced powers of self-fertilisation. Thus a professional cultivator of Cyclamens is in the habit of keeping a stock of "worthless" weedy-looking plants, for the express purpose of raising seedlings, as they are so much more prolific than the true florists' types. Having obtained them, he then crosses them, and brings them up to the standard required. Indeed, the fact is well known to all cultivators, that the poorer the plant may be, from the florists' point of view, the better seed-bearer is it; and that continually crossed and "perfect" flowers are proportionally impotent or tend to become so, when a tendency to become petaliferous often affects the essential organs.

this plant, he found that the flowers variously crossed gave the following results (omitting decimals under .6) :— *

	P.c. of Flowers.	How crossed.		Formed capsules.	Average Number of Seeds.
Long-styled	38	legit. with	mid-styled,	,,	51
,,	84	,,	short-styled	,,	107
Short-styled	83	,,	long-styled	,,	81
,,	61	,,	mid-styled	,,	65
Mid-styled	92	,,	long-styled	,,	127
,,	100	,,	short-styled	,,	108
,,	†25	illegit. ,,	long sta. of mid-st.	,,	55
,,	93	,,	long sta. of short-st.	,,	69
,,	54	,,	short sta. of long-st.	,,	47
,,	†0	,,	short sta. of mid-st.	,,	0

From these results Mr. Darwin concluded that each form of pistil is as fully fertile as possible, only when it receives pollen from the stamens of the same length as itself, these being *legitimate* unions. It will be seen that the mid-styled form is the most fertile of the three when legitimately fertilised; and as all illegitimate unions of the long- and short-styled forms were too sterile for any averages, the mid-styled form is also the most fertile when illegitimately crossed, and is least fertile with its own stamens, as indicated above by the (†). Hence self-fertilisation in this species is at a very low ebb.

A few more remarks deduced from Mr. Darwin's observations † may be added here. From the three forms occurring in approximately equal numbers in a state of nature, and from the results of sowing seed naturally produced, there is reason to belief that each form, when legitimately fertilised, reproduces all three forms in about equal numbers.

When they are illegitimately crossed with pollen *from the same form*, they evince a strong but not exclusive tendency to reproduce the parent form alone.

* *Forms of Flowers*, p. 152.   † *L.c.* p. 203.

When the short or mid-styled forms were illegitimately crossed by the long-styled, then the *two parent forms* alone were reproduced, but in no case did the third form appear.

When, however, the mid-styled form was illegitimately fertilised by the longest stamens of the short-styled, the seedlings consisted of *all three forms*. This illegitimate union was noticed as being singularly fertile, and the seedlings themselves exhibited no signs of sterility, but grew to the full height.

Finally, of the three forms, the long-styled evinces somewhat the strongest tendency to reappear amongst the offspring, whether both, or one, or neither of the parents are long-styled.

Although *L. Salicaria* has not, as far as I know, shown any signs of variability in the lengths of its filaments and styles, yet, as is perhaps generally the case with heterostyled plants, there are one or more species of the same genus which are normally homostyled. Thus *L. hyssopifolium*, which is not social, and is a dwarf form and an annual, bears only six to nine stamens, the anthers of which surround the stigma, which is included within the calyx. The three stamens, which vary in being present or absent, correspond with the six shorter stamens of *L. Salicaria*. The stigma and anthers are upturned as in the last species, and so indicate the fact that it is a degenerate form from *L. Salicaria* or some other intercrossing species, though it has now reacquired its self-fertilising properties. *Oxalis* is a genus having trimorphic species. Many of them are extremely infertile with their "own form" pollen. Such are the long-styled form of *O. tetraphylla, versicolor, Brasiliensis,* and *compressa*. On the other hand, in the long-styled form of *O. incarnata, rosea,* and *Piottæ,* and in the mid-styled form of *O. carnosa,* no self-sterility occurs.*

* According to Hildebrand, Bot. Zeitg., xlv., pp. 1, 17, 33.

ORIGIN OF HETEROSTYLISM.—The question may be now asked, How has heterostylism arisen? We have seen, in the first place, that in many cases there is a certain instability in the length of the filaments of the stamens and of the styles, in that they are liable to alter spontaneously, and especially under cultivation.* In the case of *Primula Auricula*, the homomorphic form has the anthers and stigma at the orifice, while in *P. Sinensis* they are often both low down; it is clear that either might arise in two ways. In the case of the former, the stamens, while resembling in position that of the stamens in the short-styled form, have pollen like that of the long-styled, the pistil being of that kind. Hence it is reasonable to assume that the anthers have been *uplifted*. In the Chinese Primrose it is the reverse; so that the pistil of a long-styled form has been *lowered* to the level of the stamens; the stigmas, too, are that of the short-styled.

Recognizing this instability of the essential organs, it is reasonable to assume that it may be due to varying degrees of nutrition which can readily bring about such changes, a relatively strong vegetative vigour elevating the stamens in the one case, while a slight tendency to degeneracy with lessened vital vigour tends to suppress the pistil in the other.

Assuming a homomorphic form to have been the primitive and ancestral state, we can realize how dimorphism has been brought about by such varying degrees of stimulus having been applied to the stamens and pistil. Insect agency I take to have been this cause, which, at the same time, has by selection *fixed* the heights of the stamens and style so

* See the description, given above, of *Narcissus cernuus*, Fig. 37, p. 121. Mr. Darwin found *Gilia* to vary much in this respect. It may be added that it is a not uncommon feature in flowers which are not heterostyled, as *e.g.* cultivated *Gladioli* and *Croci*, *Fritillaria Meleagris*, etc.

as to render them permanently dimorphic for legitimate fertilisation. The predominant insect or insects were (as I surmise) the direct cause of arresting the fluctuations which they themselves, as well as accidental sources of nutriment, had set up in the lengths of the essential organs, thus compelling them to retain their anthers and stigmas at the correct height.

If there were from one to three prominent kinds of insect-visitors the flowers might become adapted to them, and trimorphism be the result; if four, tetramorphism; and there is no *à priori* reason why there should not be polymorphic flowers as well, in the strict sense of the prefix of that term, provided a flower could furnish a sufficiency of stamens.

It is further to be noticed that the rule holds good with heterostyled plants, as with all other kinds of differentiation, that in nature, whenever self-fertilisation can be effected, more seed is borne than by the forms requiring intercrossing. First, whenever it can be brought about *mechanically;* as has been observed in *P. Sinensis*, by the corolla, when falling off, dragging the anthers over the stigma in the long-styled form, which consequently yields more seed.* In *P. veris*, it does not do so; but as pollen can *fall* in the short-styled form, in this species that form is thus the most fertile (see above, p. 205).

Secondly, when these plants are artificially and legitimately fertilised, and not left to the chance visits of capricious insects, then the results are as they should be; but if self-fertilisation be artificially and *repeatedly* practised, then nature responds to the act; the anthers and pollen may in part degenerate, but what is *left good* is ample to secure abundant seed, and the self-fertilised form surpasses even the

* Darwin found that, in the absence of insects, the long-styled form of *P. Sinensis* was twenty-four times as productive as the short-styled.

legitimately fertilised heteromorphic unions in fertility. Thus, Mr. Darwin observes, "The self-fertility of *Primula veris* increased after several generations of illegitimate fertilisation, which is a process closely analogous to self-fertilisation." *

Lastly, if homomorphic forms occur spontaneously, as is often the case with species of *Primula*, Mr. Darwin has shown they are not only "capable of spontaneous legitimate fertilisation, but are rather more productive than ordinary flowers legitimately fertilised." †

It was Mr. Scott who suggested that the equal-styled varieties arose through reversion to a former homostyled condition of the genus. Mr. Darwin supported this view in consequence of observing "the remarkable fidelity with which the equal-styled variation is transmitted after it has once appeared." ‡

* *Cross and Self Fertilisation*, p. 351.
† *Forms of Flowers*, p. 273; and *Cross and Self Fertilisation*, p. 352.
‡ *Forms*, etc., p. 274; Mr. Darwin was so profoundly impressed with the supposed advantages of intercrossing, that he again and again asserts most positively that self-fertilisation is injurious, often in diametrical opposition to his own statements and experiments. Thus, while speaking of heterostyled trimorphic plants, he says, "As I have elsewhere shown (*The Effects of Cross*, etc.), most plants, when fertilised with their own pollen, or that from the same plant, are in some degree sterile, and the seedlings raised from such unions are likewise in some degree sterile, dwarfed, and feeble." Yet, in the work quoted, he has not only shown that, when he persevered with self-fertilisation for several generations, he found it was just the reverse; as *e.g.* with "Hero" *Ipomœa*, the white *Mimulus*, etc., and with *Primula*, as stated above; but he more than once draws an opposite conclusion, as when speaking of self-fertile varieties (*l.c.*, p. 352): "It is difficult to avoid the suspicion that self-fertilisation is in some respects advantageous.... Should this suspicion be hereafter verified, it would throw light on the existence [of cleistogamy]." It is this "suspicion" which I have completely verified; and, indeed, any idea of "injuriousness" is refuted by the *majority*

Besides the more obvious differences in the relative lengths of the styles and filaments* of heterostyled flowers, the rule is for the stigmas of the long-styled to be larger or longer than those of the short-styled,† and to have their papillæ longer and broader.

Thus in the nine species of *Primula* described by Mr Darwin, in two only were the stigmas nearly alike in both. Of three species of *Linum*, *L. flavum* alone had an appreciable difference in the stigmas. In *Pulmonaria officinalis* and *Polygonum fagopyrum*, *Forsythia suspensa* and *Ægiphila elata*, it was not, or scarcely appreciable.

Again, besides those mentioned there were twenty species in which the stigmas of the long-styled were markedly superior to those of the short-styled.

of plants in a wild state being constantly self-fertilised, as Müller, and, indeed, Mr. Darwin himself has shown to be the case. Thus, he gives two lists, of forty-nine species in each, (*Cross and Self Fert.*, etc., pp. 357 and 365), one of self-sterile, the other of self-fertile plants, and adds, " I do not, however, believe that if all known plants were tried in the same manner, half would be found to be sterile within the specified limits; for many flowers were selected for experiment which presented some remarkable structure ; and such flowers often require insect aid " (*l.c.*, p. 270). The proportion of self-sterile plants is, in fact, extremely small. Müller remarks, *e.g.*, of the highly differentiated order *Scrophularineæ*, that " in default of insect-visitors, self-fertilisation takes place in most forms; and in only a few are insect-visits, and consequently cross-fertilisation, so far insured that self-fertilisation is never required and has become impossible." Similarly of *Labiatæ* he says, " Self-fertilisation seems to be rendered impossible only in the species of *Nepeta*, *Thymus*, *Mentha*, and *Salvia* described " (*Fertilisation*, etc., pp. 464 and 503). Moreover, while Mr. Darwin includes the Fox-glove and *Linaria vulgaris* among his sterile plants, Müller considers them both to be self-fertilising.

* Exceptions occur, thus *Cordia* and *Linum grandiflorum* have little or no difference in the length of the stamens.

† *Leucosmia Burnettiana* is remarkable for having the stigma of the short-styled form the more papillose (*Forms of Flowers*, p. 114).

On the other hand, the anthers of the short-styled are usually longer and contain larger pollen grains than those of the long-styled, the pollen of which is also often more translucent and smoother.

Of all the species included in the above-mentioned thirty-six species, only five seem to have the pollen of both forms of the same size, and two in which it was reversed. The five species are *Leucosmia Burnettiana, Linum grandiflorum, Cordia, Gilia pulchella,* and *Coccocypselum.* The two in which the pollen grains of the long-styled form were the larger, were *Gilia micrantha* and *Phlox subulata.*

The presence of cases where the usual differences are not pronounced is just what one expects to find, in accordance with the laws of differentiation; whereby intermediate conditions are to be looked for. Thus some species of *Primula* afford great differences in the shapes of the stigmas, *P. veris* being globular in the long-styled, and depressed in the short-styled; while in *P. Sinensis* it is elongated: but in other species, as *P. Sikkimensis* and *P. farinosa,* there is but little difference between the stigmas of the two forms. In some cases the differences reside entirely in the stamens or pollen grains, as in *Forsythia suspensa,* in which, although (contrary to the rule) the anthers of the long-styled are in length as 100 : 87 compared with the short-styled, yet the pollen grains are as 94 : 100, which agrees with the rule. With *Linum grandiflorum* and *Cordia* and *Gilia pulchella,* etc., the difference lies in the pistil. On the other hand, the difference may reside in the stamens, as in *Ægiphila elata,* the pollen grains being as 62 : 100, *i.e.* in the long-styled as compared with the short-styled.

Ægiphila obdurata has the stigmas of the long-styled in length 100 : 55 as compared with the short-styled; and the length of the anthers as 44 : 100. This is, therefore, apparently truly heterostyled, but from Mr. Darwin's observations he thinks the short-styled incapable of fertilisation; moreover the anthers of the long-styled form were " brown, tough, and devoid of pollen." He considers that, from having been heterostyled, it has now become diœcious, or else gynodiœcious.

M. W. Burck has shown * that several genera of *Rubiaceæ* are heterostyled in form but quite diœcious.

Faramea affords another curious difference. In the long-styled form the stigma is short and broad; in the short-styled, it is long, thin, and curled. The anthers of the short-styled are a little larger than those of the long-styled, and the size of their pollen grains are as 100 : 67. But the most remarkable difference (of which no other instance is known) is in the fact that while the pollen grains of the short-styled forms are covered with sharp points, the smaller ones are quite smooth. The anthers, moreover, rotate outwards in the short-styled, but do not do so in the long-styled flowers. A similar rotation takes place in some of the *Cruciferæ*, and facilitates intercrossing. A somewhat analogous torsion occurs in some styles and stigmas, as of *Linum perenne*, *Luzula arvensis*, *Begonia*, etc.

The smaller and smooth pollen, in the more degenerate condition of the long-styled form, is suggestive of the origin of that of wind-fertilised flowers, which has sometimes acquired the same form. Indeed, the two forms of pollen (figured by Mr. Darwin at p. 129 of *Forms of Flowers*) exactly correspond to the very common spinescent form in intercrossing species of *Compositæ*, and to that of the anemophilous *Artemisia* of the same order, respectively.

The general conclusion, therefore, derived from the com-

* *Sur l'Organisation Florale chez quelques Rubiacées.* Ann. Jard. Bot. Buitenzorg 3, p. 105.

parison of these minute details, is that the long-styled form of flower represents a more fully developed pistil, and therefore a more female condition; while the short-styled is more male: and, as we have seen above, this is borne out by the comparison of the offspring; and, lastly, by the probable diœcious condition of *Ægiphila obdurata*, as well as by the actual diœcism of some species of *Mussænda* and *Morinda umbellata;* while *Mussænda cylindrocarpa* and certain other species of *Morinda* are hermaphrodite without heterostylism (Burck, *l.c.*).

CHAPTER XXIV.

PARTIAL DICLINISM.

GYNODIŒCISM AND GYNOMONŒCISM.*—In accounting for the origin of certain floral structures, it must be borne in mind that the habits and constitutions of plants are so infinitely various, that the interpretation given for that of a structure in one case may fail to be satisfactory when tested by another; and an argument apparently sound for the explanation of a special phenomenon in a particular plant or plants may not at all apply to that of others. Thus, while the Hazel may mature its stamens before the pistils on a slight rise of temperature in early spring, there are many herbs, if they happen to blossom in spring earlier than is their custom, in summer, or what may be their *optimum* period, may have the staminal whorl more or less deranged, as such plants require a relatively higher temperature to develop them perfectly.† This is particularly characteristic of gynodiœcious plants. Thus, *e.g.*, most of the distinctly protandrous species of the *Alsineæ* are in this condition, and

---

* *Gynodiœcism* signifies that the same species may have both female and hermaphrodite plants.

Gynomonœcism signifies that the same plant may bear both female and hermaphrodite flowers.

† This will be discussed more fully in the next chapter.

the plants with small, usually pistillate flowers are chiefly in blossom at the beginning of the flowering period of the larger-flowered hermaphrodite plants of this section of the *Caryophylleæ*. Similarly, *Caffea arabica* produces small pistillate flowers in Guatemala at the beginning of the season.* It is the same with *Geranium macrorhizon* and many species of *Pelargonium*, etc. †

Gynodiœcism also prevails in the *Labiatæ*, but both female and hermaphrodite plants for the most part blossom simultaneously in summer. It may be noticed that the corolla is almost invariably reduced in size in female flowers, whether the species be strictly diœcious as in Bryony, or gynodiœcious as Thyme, showing the close interdependence between the corolla and stamens. ‡

That climatal conditions are likewise connected with the Gynodiœcism of the *Labiatæ* seems probable from the behaviour of *Thymus Serpyllum*; for Delpino found that it was trimorphic in the warmer region of Florence, having flowers with greatly developed stamens and the pistil in every stage of abortion or even absent (see Chapter XXV.); other flowers showed the exact converse; and, lastly, others were hermaphrodite. Müller, however, on the other hand, in Westphalia and Thuringia; Ascherson, in Brandenburg; Hildebrand, in the Rhine provinces; and Mr. Darwin, in England, never met with the purely male form; though Dr. Ogle found some with the pistil permanently immature.§ Similarly, *Eriophorum angustifolium* is gynodiœcious in Scotland and the Arctic regions.‖

Besides temperature, the character of the soil has most probably much effect in bringing about this kind of partial

* Müller, *Fertilisation*, etc., p. 304. † *L.c.*, p. 158.
‡ See *Forms of Flowers*, pp. 307–309.
§ Müller, *l.c.*, p. 474. ‖ *Forms of Flowers*, p. 307.

diclinism. Mr. Darwin thought "a very dry station apparently favours the presence of the female form,"* *i.e.* a lessened vegetative vigour tends to check the development of the corolla and stamens, especially if a low temperature accompanies it; just as, conversely, we have seen how a high temperature enhances it. Mr. Hart thus found that, with *Nepeta Glechoma*, all the plants which he examined near Kilkenny were females; while all near Bath were hermaphrodites, and near Hertford both forms were present, but with a preponderance of hermaphrodites.†

Both Müller and Mr. Darwin offer theories to account for the origin of these gynodiœcious plants.

Müller, after quoting Hildebrand's view, which he rejects,‡ says,§ "Of the flowers of the same species growing together, the most conspicuous are first visited by insects, and if the flowers on some plants are smaller than on others, perhaps owing to scanty nourishment, they will generally be visited last. If the plant is so much visited by insects that cross-fertilisation is fully insured by means of protandrous dichogamy, and self-fertilisation is thus rendered quite needless, then the stamens of the last-visited small-flowered plants are useless, and Natural Selection will tend to make them disappear, because the loss of useless organs is manifestly advantageous for every organism.

"This explanation rests upon the hypotheses, (1) that the flowers of those species in which small-flowered female plants occur together with large-flowered hermaphrodite plants are plentifully visited by insects and are markedly

* *Forms of Flowers*, p. 301
† *Nature*, 1873, p. 162; and see below, p. 239.
‡ *Fertilisation*, etc., p. 473.
§ *L.c.*, p. 484. Compare his remarks on *Scabiosa arvensis*, *l.c.*, pp. 310, 311.

protandrous; (2) that variation in size of the flowers has always taken place, not among the flowers on a single plant, but between the flowers on different individuals."

Mr. Darwin suggests another view:* "As the production of a large supply of seeds evidently is of high importance to many plants, and . . . the females produce many more seeds than the hermaphrodites, increased fertility seems to me the more probable cause of the formation and separation of the two forms."

"S. M.," reviewing Mr. Darwin's work in the *Journal of Botany*, 1877, p. 375, "felt compelled to differ from the author, and adds, "For ourselves we cannot help thinking that gynodiœcism can be better explained on the view of a sufficiency of pollen for the fertilisation of all the individuals of a species being produced by only a few of the flowers, so that instead of some of the anthers of all the flowers becoming abortive—a very common occurrence—we see here abortion of all the anthers of some of the flowers. . . . All known instances of gynodiœcism relate to species which have the maximum of stamens possessed by the orders to which they relatively belong, and are without any complex entomophilous structure. . . . We may also remark on the panciovulate condition of gynodiœcious species, and ask why do we not see this form of sexual separation in multiovulate ones?"

In reply to this writer's suggestions, I would remark that in *all* entomophilous flowers far too much pollen is produced and wasted; that Mr. Darwin's observation, that a bee could fertilise ten pistils with pollen from one flower of *Satureia*, might readily apply to hundreds of cases where no gynodiœcism exists; and as long as insects visit flowers the tendency is *not* to contabescence and abortion of the

* *Forms of Flowers*, p. 304.

anthers, but to higher differentiations and an increase in the quantity of pollen. Secondly, that the orders, with gynodiœcism have the maximum of stamens, is not universally true, *Pelargonium* having only seven out of ten. Again, the *Labiatæ* are especially characterized by "entomophilous structures." Lastly, the order *Caryophylleæ* is multiovulate.

In the first two interpretations, those of Müller and Darwin, Müller suggests scanty nourishment as a cause for the diminished size of the female flowers, which might apply to any or every protandrous plant and so give rise to gynodiœcism; for if it be a sufficient cause in one family, why has it not brought it about in all? This cause alone does not touch the question, Why is gynodiœcism peculiarly common in the *Alsineæ* of the *Caryophylleæ* and in *Labiatæ?* Mr. Darwin thinks that an increased fertility of the female may be the cause; but he seems to forget that no flower of the *Labiatæ* can bear more than four seeds, so that, supposing a female plant to have the same number of flowers as a hermaphrodite, if it bears more seeds it must be due to *the decrease in fertility* of the latter, and not to any increase in the former.* It is, in fact, a very common occurrence for a flower of any member of the *Labiatæ* to bear one, two, or three only, as well as four nutlets in an individual fruit. Mr. Darwin "doubts much whether natural selection has come into play," and notices that "the abortion of the stamens ought in the females to have added, through the law of compensation, to the size of the corolla," as is the case in the ray florets of the gynomonœcious *Compositæ*. He, however, recognizes the

---

* In his experiment with *Satureia hortensis*, Mr. Darwin collected seeds from the finest of ten female plants, and they weighed 78 grains; while those from the single hermaphrodite, which was a rather larger plant than the female, weighed only 33·2 grains; that is, in the ratio of 100 to 43 (*Forms of Flowers*, p. 303).

PARTIAL DICLINISM. 225

intimate connection between the corolla and andrœcium, and thinks that "the decreased size of the female corollas is due to a tendency to abortion spreading from the stamens to the petals."

In noting all the plants mentioned by Müller and Mr. Darwin as gynodiœcious, there are besides the two well-marked groups already mentioned, viz., *Alsineæ* and *Labiatæ*, the following isolated genera or species, *Pelargonium, Geranium macrorhizon, Sherardia arvensis, Valeriana montana, Scabiosa, Cnicus, Echium vulgare* and *Plantago;* to the *Compositæ*, I can add *Achillæa millefolium;* and I think also Vines may be included in the list.

The first and important point to note about the flowering of the *Alsineæ* is that the female flowers *are the first to open, at the beginning of the season.** It is the same with *Geranium macrohizon, Pelargonium,* and Coffee in Guatemala. Now, we have already seen how sensitive the andrœcium and the corolla are to a low temperature, so that we have here a direct cause which will account for the check upon the growth and development of these two whorls. Applying this principle to the *Labiatæ*, we must remember that as a group they are correlated to a warmer climate, their "home" being the Mediterranean and even warmer regions; hence I assume their greater hereditary sensitiveness to a low temperature in those descendants which occupy a cooler temperate zone. This may, I think, account for the predominance of purely female forms, as well as the presence of stamens in every degree of degeneracy.

How far the same principle will apply to the other gynodiœcious genera and species, I will not pretend to offer an opinion, as not enough is yet known about them;

* See Hildebrand's observation, p. 234, and SEXUALITY AND TEMPERATURE, p. 237.

only we must always remember that there may be a variety of causes which may equally well bring about the same result.

It may be also borne in mind here that another result of low temperature is, while retaining the function of the androecium, to arrest the expansion of the corolla and to render the flowers self-fertilising. This is peculiarly the case with the *Alsineæ*; while *Lamium amplexicaule* fails to open its earliest small-flowered flowers at all, being strictly cleistogamous.

The preceding cases of gynodicecism are all associated with a more or less degree of protandry. It is rarer to find it accompanied with protogyny in the hermaphrodite form. Müller records it in *Plantago lanceolata* in England, which I can corroborate, and in *P. media* in Germany. These plants are anemophilous, and in a state of passage from an entomophilous ancestry; so that it may have been retained from an early condition.

Gynomonœcism is not particularly common, except in the *Compositæ*, where the ray florets are often female, while the disk florets are hermaphrodite. This is due to compensation; for transitional states may be seen in flowers which are passing into the "double" condition; for as the corolla changes its form and becomes ligulate, the stamens are suppressed, and the style arms alter their shape. *Anemone hepatica* is said to be gynomonœcious,[*] and also *Syringa Persica*.[†] I have seen no case, and no description is given of these two, so that I can only suggest that it may be a result from degeneracy, perhaps on the road to a petaloid condition of the stamens. Such a state I have found in a *Plantago* which was gynodicecious.

[*] Dr. S. Calloni, *Arch. Sci. Phys. et Nat.*, xiii., 1885, p. 409.
[†] Müller, *Fertilisation*, etc., p. 393.

ANDRODIŒCISM AND ANDROMONŒCISM.*—These conditions do not appear to prevail to the same extent as the female forms of flowers. Both of these kinds are not at all uncommon in the *Umbelliferæ*, and are a result of exhaustion, for the umbels produced at the end of the season are often entirely male; or, if at other periods, it is generally the central florets which develop no pistils, as in *Astrantia minor*. Müller has noticed how "the weaker plants usually bear but one umbel consisting only of male flowers." This would make it androdiœcious. I find that andromonœcism prevails in *Astrantia major, Carum, Smyrnium*, and in *Trinia vulgaris*. This last, growing on the Clifton downs, bore umbels which were altogether male, *after* the hermaphrodite ones had formed their fruit. *Daucus grandiflora* is remarkable for having three kinds of flowers. According to Müller, the central ones are male; at the edge of the umbellule the flowers are neuter, with the outermost petal greatly enlarged; lastly, at the margin of the whole umbel, are female florets in which the outer petals attain to a gigantic size.†

* *Androdiœcism* signifies that the same species has both male and hermaphrodite plants.
Andromonœcism signifies that the same plant bears both male and hermaphrodite flowers.

† I would here remind the reader that the interpretation given above (Chapters XI.-XIII.) of the origin of irregular corollas, applies equally well to those cases where it is only in the outermost florets of a cluster where the petals are enlarged, as in *Iberis*, many of the *Compositæ*, and *Umbelliferæ*, as well as in *Hydrangea*, Guelder Rose, etc. In all these, when insects first approach the umbel and alight on the border of it, any or each individual floret on the margin may have to carry the burden. As soon, however, as the insect passes the edge of the cluster, its weight is distributed over several florets; so that they are not submitted to any special strains upon one, *i.e.* the outer side only. The same remarks apply to *Mentha*, as compared with *Lamium*. The insect visits one flower at a time in the latter, but scrambles over several in the former, which has (presumably) degraded in consequence.

Caltha palustris is said to be androdiœcious, but no details are given by the observer.*

Besides the *Umbelliferæ*,† where andromonœcism seems to be a characteristic feature, Müller mentions *Asperula taurina* and *Galium Cruciata, Pulmonaria officinalis, Coriaria myrtifolia,* and *Diospyrus Virginiana* as being andromonœcious. The hermaphrodite flowers of these species are protandrous.

In *Galium Cruciata*, Mr. Darwin noticed that the pistil is suppressed in most of the lower flowers, the upper remaining hermaphrodite.

Heterostylism may tend to produce the same result when the stamens of the long-styled forms degenerate so far as to become atrophied without the pistil losing its functions. *Pulmonaria angustifolia* and *Phlox subulata* give hints of this condition.‡ *Asperula scoparia* was at first thought by Mr. Darwin to be heterostyled, but finding the anthers to be destitute of pollen, he considered it to be diœcious. *A. taurina*, as figured by Müller,§ shows great variability in the lengths of the filaments and styles, and he pronounces it to be andromonœcious. Hence, as so many of the *Rubiaceæ* are heterostyled, there seems every probability of one result of this peculiarity, being one or other kind of this incompletely affected or partial diclinism. In the case of *Coriaria myrtifolia*, Hildebrand found that it was the first flowers which were male only. In Maples, as in *Galium Cruciata*, the rule is for the three or more flowered corymb to have the central one hermaphrodite, and the lower or outer ones male. This

---

* Lecoq, *Geóg. Bot.*, tom. iv., p. 488.

† Müller says that in *Sanicula Europœa* the *outer* flowers are male, and develop *after* the inner ones, which are hermaphrodite. This is so anomalous, that one suspects an error somewhere. I have not had any opportunity of examining fresh flowers.

‡ *Forms of Flowers*, p. 287.

§ *Fertilisation*, etc., p. 303.

PARTIAL DICLINISM.

clearly is a question of the distribution of nutrition; the *lower*, being the *later* ones to expand, are the weaker.* Müller mentions Horse-chestnuts as being also andromonœcious; and what is exceptional is that the hermaphrodite flowers are protogynous. This, however, may be due to the early period of flowering, like species of *Prunus* and *Cratægus*.

The reader will now perceive that there may be several causes at work to produce these kinds of "partial diclinism;" and that what is required is to ascertain, if possible, by observation and experiment, which is the one peculiar to each species. Secondly, when any one or more causes has been sufficiently persistent, the results become hereditary; so that certain species, genera, and orders become more or less characterized by these peculiar features.

* Compare the observations on *Adoxa*, p. 188.

CHAPTER XXV.

SEXUALITY AND THE ENVIRONMENT.

GENERAL OBSERVATIONS.—As the environment is now known to have most potent influences on the anatomical structure of the vegetative system of plants, thereby affecting their outward and visible morphological characters as well; so are there many causes which affect the reproductive system, at one time influencing the androecium, at another the gynœcium, favouring them or the reverse as the case may be; so that either sex or even both may be entirely suppressed, and a hermaphrodite flower become male, female, or neuter.

With regard to the most general agency, there seems to be a tolerably uniform consensus of opinion that the female sex in plants is correlated with a relatively stronger vital vigour than the male; and this is just what an *à priori* assumption would look for, as the duration of existence and the work to be done in making fruit require a greater expenditure of energy than the temporary function of the stamens.

We must, however, distinguish between a healthy *vital* vigour, and any excessive vegetative growth, as occurs under high cultivation, and as is often the result of intercrossing. If this latter surpass the requisite or optimum conditions for the healthy performance of the functions of all the organs of a plant, then either of the sexual organs may begin to deterioriate, till they become metamorphosed into petals or leaves, or else degenerate and vanish.

It is true enough that we know nothing of the real nature of life; but it is easy to see that, of the various phases of development, from germination to the production of seed, each should have the proper amount of energy at its disposal, and no more; for if any one organ be stimulated beyond the optimum degree, others suffer through atrophy. The first and well-known distinction to be noticed lies, of course, between the "vegetative energy," by means of which roots, stems, branches, and foliage are developed, and the "reproductive energy," which brings about the formation of flowers, fruit, and seed. If either of these be unduly excited, the other diminishes. Thus, as long as fruit trees are developing much wood and foliage, they either bear fruit badly or not at all. Plants which are propagated largely by vegetative means of multiplication, such as bulbs, corms, tubers, etc., are notorious for failing to set seed as well. As an instance in nature, *Ranunculus Ficaria* may be mentioned. This plant propagates itself by "root-tubers" and by aërial corms, and rarely produces much fruit, for the pollen often remains in an arrested state.* Conversely, if vegetative energy be checked by root and branch pruning, bark-ringing, etc., the reproductive energy is promoted, and an abundance of fruit is the reward. Similar results follow a decrease of energy through impoverishment, when enormous crops of fruit may be borne by trees, as I have seen in Portugal Laurels, when the roots had penetrated a bed of gravel and the branches became decayed.

Apart from these general considerations certain special conditions are found to favour one sex more than the other,

* See Van Tieghem on R. Ficaria, *Ann. des Sci. Nat.*, v., sér. 5, p. 88.

so that normally hermaphrodite flowers may become unisexual, and every possible degree between these two extreme cases can be met with in nature and cultivation. The problem, therefore, is to discover what the immediate causes may be in each case which stimulate or suppress the energy required for the proper development of the stamens and pistil respectively.

There appears to be a closer bond between the stamens and corolla than between the two kinds of essential organs themselves; * thus, if the corolla degenerate, the antipetalous stamens at least tend to follow suit, as in the *Alsineæ*. On the other hand, the first tendency towards "doubling" appears in a more or less pronounced petalody of the andrœcium.

As petals are a nearer approximation to foliar organs, the above means that vegetative energy is more prone to affect the stamens, when from some cause they have first begun to lose their proper function, than the pistil.

The pistil may fail in its development from two classes of causes: either from an undue display of the vegetative vigour, as in *completely* double flowers—though it may be unaffected in a *partially* double one; or else from excessive feebleness, under which a flower may succeed in making the andrœcium, but has not sufficient energy to develop the gynœcium; as, *e.g.*, often takes place in the flowers of the *Umbelliferæ* at the close of the season.

There is no absolute rule in these matters, and differences result from various degrees of energy at the disposal of the

---

* A study of the vascular system of flowers and their axes bears this out, as the provision made for the stamens usually arises from the perianthial cords, while that for the pistil is mostly isolated off in rather a more marked and independent manner. Exceptions occur, as in *Ballota nigra*, in which the four stamens originate from the same cords as those of the placentas.

whorls, giving rise to corresponding results of different degrees of development in the respective sexes.

The points to be clearly perceived are that a plant should be able to develop all its organs in perfection; that there is an optimum degree of energy for each; and that, though it is customary to group these energies under the two expressions, vegetative and reproductive, yet the principle may be carried out in detail: so that, *e.g.*, an enlarged corolla tends to destroy the stamens, as of the ray florets of *Dahlia*, or even the pistil too, if it be very large, as in *Centaurea*. A stimulated andrœcium brings about an arrest in the pistil, and causes protandry; and if the perianth be highly developed, as in orchids, the enhancement of the former may cause degeneracy in the ovules.

SEXUALITY AND NUTRITION.—Assuming, for the present, that the ancestral condition of all flowers, excepting, perhaps, those of the Gymnosperms, was hermaphrodite, many instances exist of the same species having male, female, and hermaphrodite flowers, such as the Ash, *Silene inflata*, etc., where the aborted organs often remain more or less rudimentary. It cannot be pretended as yet that the cause or causes can be at all positively asserted, in each case, for the tendency to abortion either in the stamens or pistil; but there are certain well-ascertained facts which can undoubtedly play a part in the processes of degeneration or exaltation of the staminal and carpellary energies respectively. If they be sufficiently persistent the subsequent generations can, then, become completely diclinous, without a trace of the other sex remaining; yet, as is well-known, any diclinous plant may reproduce by reversion the lost sex, thereby revealing its original hermaphroditism.

In endeavouring to trace the present condition of diclinous flowers back to an ancestral hermaphrodite condition, it will be as well to consider certain significant facts which may help us in ascertaining the cause of their present diclinism.

"Hildebrand has shown," writes Mr. Darwin, "that with hermaphrodite plants which are strongly protandrous the stamens in the flowers which open first sometimes abort. . . . Conversely the pistils in the flowers which open last sometimes abort." Similarly Gärtner observed that "if the anthers on a plant are contabescent (and when this occurs it is always at a very early period of growth) the female organs are sometimes precociously developed." *

A reason for this is that, on the one hand, since a higher temperature is correlated with protandry, the first flowers open when the optimum temperature has not arisen; so that the stamens are checked, a cooler temperature being less inimical to the development of the gynœcium. On the other hand, the last flowers of the season are produced when the vital energy is waning, and although the flowers may expand, they are too feeble to develop the pistil.

Now exactly the converse may occur; thus Mr. W. G. Smith called attention † to the seemingly unobserved fact that *Euphorbia amygdaloides* always bears terminal male flowers alone at first, and subsequently the two sexes together on lower lateral "flowers." This agrees with *Castanea Americana*,‡ as noticed by Mr. Meehan. In these two cases,

* *Forms of Flowers*, p. 283. I hardly think this can be always the case; for, of Vines growing side by side, some will occasionally have the anthers utterly devoid of sound pollen, but with the pistil normal; while others will be entirely hermaphrodite with no sign of contabescence. I have examined such, supplied to me by Mr. Barron from the gardens of the Royal Horticultural Society at Chiswick. The cause is at present very obscure.

† *Journ. of Bot.*, 1864, p. 196.

‡ *Proc. Acad. N. Sci. of Philadel.*, 1873, p. 290.

therefore, we have instances of the plants flowering and bearing male organs only *before* the highest effort of vital energy is displayed—the preliminary and feebler effort being capable of developing the andrœcium alone.

With regard to diclinous trees, many examples could be found to illustrate the principle that the female flowers are normally produced by stronger shoots than the male. Mr. Meehan has particularly called attention to this fact. For instance, "*Juglans nigra** exhibits three grades of growing buds. The largest make the most vigorous shoots. These seem to be wholly devoted to the increase of the woody system of the tree. Lower down, the strong last year's shoots arise from buds not quite so large. These make shoots less vigorous than the other class, and bear the fema'e flowers on their apices. Below these are numerous small weak buds, which either do not push into growth at all, or when they do, bear simply the male catkins."

Again, *Castanea Americana* bears two crops of male flowers, the first of which disarticulate and are useless; the second appear about ten days later, accompanied by clusters of females. Occasionally a tree will be entirely female.

Mr. Meehan also calls attention to the fact that isolated trees of Birch, though producing an abundance of male and female flowers, very often have not a perfect seed. Hazels are sometimes protogynous, sometimes protandrous; and if the latter condition prevail, there may be little or no fruit, as often occurs in Pennsylvania. After making analogous observations on American Maples, he summarizes his remarks on the latter as follows:—

"Male flowers do not appear on a female Maple-tree till some of its vital power has been exhausted.

* *Laws of Sex in J. Nigra*, Proc. Acad. N. Sci. of Phil., 1873, p. 290.

"Branch-buds bearing female flowers have vital power sufficient to develop into branches.

"Branch-buds bearing male flowers have not vital power enough to develop into branches, but remain as spurs, which ever after produce male flowers only.

"Buds producing male flowers only, are more excited by a slight rise of temperature than females, and expand at a low temperature under which the females remain quiescent" [*i.e.* when the winter temperature begins to give way to the rise in early spring, the males are more easily excited into maturity]. *

As another authority, I would refer to a paper by Mr. Moore, upon the appearance of male flowers on female trees, such as the Papaw, etc. He alludes to Dr. Wight's views, in that he attributes these changes "to the modifying power of the soil and climate acting on the dormant energies of the rudimentary ovaries and developing them into prolific fruit, but at the cost of the male organs." In another case of the Papaw one fertile flower was produced, and that the first which expanded, others being all male "It would seem that fertile flowers in these instances have only been developed when the greatest vital energy is present in the plant, which is the case when they first begin to expand. Other instances," Mr. Moore adds, "might be quoted to show that vigour and healthiness increase the female line of vital force in vegetables, whilst weakness is more conducive to the male development."

This view was corroborated by a case of a young plant of *Nepenthes distillatoria*, raised from seed. Mr. Moore describes and figures it in the same paper. The lowermost flowers of the raceme bore both stamens and pistil, the

---

* *On the Relation of Heat to the Sexes of Flowers*, Proc. Acad. Nat. Sci. of Phil., 1882, p. 1.

carpels of which were somewhat dissociated. On the upper half they were entirely male. He did not succeed in impregnating any of the numerous and well-formed ovules. He observes: "This well-authenticated case also favours the theory that vigour in the plant is productive of the female line of vital force." *

It is a common phenomenon for diclinous trees to change their sex in different places or seasons. Ashes and Maples, as well as Palms, have been known to do this. The only interpretation being apparently the difference which occurs in the climatal conditions from year to year, or the modifications of temperature, soil, etc., consequent on different environing circumstances.

SEXUALITY AND TEMPERATURE.—Temperature has a marked influence on the sexes. A relatively high temperature favours the corolla and andrœcium, while a comparatively lower one the gynœcium. A. Knight long ago found that Watermelons grown with a maximum of 110° by day, usually varying from 90° to 105°, with a minimum of 70° at night, grew with luxuriance, but bore no fruit, though it had a profusion of minute male blossoms. This experience is corroborated by present horticulturists. He was not surprised, as he had for many years previously succeeded, by long-continued low temperature, in making cucumber plants produce female flowers only.

Mr. Meehan's observations on the development of buds on certain trees appeared to corroborate this view of Knight's. He remarks that, in the year 1884, after a winter of uniformly low temperature, the male and female flowers of the nut appeared together; but in other years it was

* *Trans. Irish Acad.*, xxiv., p. 629; see also a paper on "Sexuality," by Dr. M. T. Masters, *Pop. Sci. Rev.*, xii., p. 363, 1873, and his *Teratology*, p. 190; also, *Proc. Acad. Nat. Sci. of Phil.*, 1873, p. 290.

found that a few warm days in winter would advance the male flowers, so that they would mature some weeks before the female flowers opened. Hence the latter were generally unfertilised.*

That the stamens are much more sensitive to and precocious in their development under a rise of temperature, is seen in the behaviour of plants in different countries. Thus it is asserted † that *Stratiotes aloides* produces its carpels with greater abundance towards the northern limit of its geographical distribution, and its stamens, on the contrary, are more frequently developed in more southern districts. ‡

These tendencies to check one or the other sex, may lead to monœcious diclinism; and even complete diœcism seems, at all events to some extent, due to climate, as differences occur in widely separated countries; thus *Honchenya peploides* is frequently hermaphrodite in America, but usually sub-diœcious in England.§

Mr. Darwin, in his experiments, found that *Mimulus luteus* was very sterile in one year; and he attributed the fact partly to the extreme heat of the season.‖

* *Proc. Acad. Nat. Sci. of Phil.*, 1884, p. 116.
† *Teratology*, p. 196.
‡ Perhaps the propagation by apogamy of the female plants of *Chara crinita* may be a resource to which this plant has been driven in consequence of the male plants not thriving in a cool region. Sachs says that the female is found throughout the whole of Northern Europe, but the male is only known to occur in Transylvania, South of France, and by the Caspian (*Phys. of Plants*, p. 801).

The idea is suggested by this that when temperature arrests the male without checking the vegetative system, a plant may adopt vegetative methods of multiplication. Thus, instead of regarding the "root-tubers" and aërial corms of *Ranunculus Ficaria* as the cause of the degeneracy of the pollen in that plant; perhaps it would be more correct to reverse the process.

§ *Teratology*, p. 196.   ‖ *Cross and Self Fert.*, etc., p. 68.

Mr. Darwin also records * how "a tendency to the separation of the sexes in the cultivated Strawberry seems to be much more strongly marked in the United States than in Europe; and this appears to be the result of the direct action of climate on the reproductive organs." Quoting from the *Gardener's Chronicle*,† he adds, "Many of the varieties in the United States consist of three forms, namely, females, which produce a heavy crop of fruit,—of hermaphrodites, which 'seldom produce other than a very scanty crop of inferior and imperfect berries,'—and of males which produce none.... The males bear large, the hermaphrodites mid-sized, and the females small flowers. The latter plants produce few runners, whilst the two other forms produce many; ... we may therefore infer that much more vital force is expended in the production of ovules and fruit than in the production of pollen."

Conversely, as runners were more abundant with male and hermaphrodite plants, we see here an instance of *vegetative growth* correlated with the male elements at the expense of the female.

SEXUALITY AND THE SOIL.—Müller has given two instructive cases where it is pretty certain that the soil was a chief cause of the separation of the sexes.‡ *Dianthus deltoides*, near Lippstadt, offers interesting gradations from hermaphroditism to gynodiœcism and gynomonœcism. "On the border of a meadow, of some hundred stems examined by myself, all the flowers, without exception, proved to be protandrous, with a normal development of the anthers and stigmas. On the grass-grown slope of a sandy hill likewise, all the stems produced protandrous flowers, but on many stems the stamens, although emerging above the petals

* *Forms of Flowers*, p. 293.   † 1861, p. 716.
‡ *Nature*, vol. xxiv., p. 532.

before the development of the styles and stigmas, bore diminished whitish anthers, not opening at all, and containing also some shrivelled pollen-grains. Lastly, in a barren sandy locality, many of the stems produced female flowers, with stamens aborted in the same degree as in *D. superbus*, and not infrequently such female flowers and protandrous hermaphrodite ones are found on the same stem." Wiegman also found the *Dianthus* had contabescent stamens when growing on a dry and sterile bank. The conditions here mentioned are very like those more than once described as associated with double flowers, in which the stamens have also degenerated but taken the petaloid form. Hence I think we may directly trace the degeneracy of the anthers and pollen to atrophy; since chemical analyses of pollen prove that the most important constitnents required are potash, nitrogen, and phosphorus pentoxide,[*] probably wanting in the localities mentioned.

"*Centaurea Jacea*" Müller describes[†] "as having its flower-heads of the same stem always of the same form, but different stems of the same locality often present astonishing differences in their flower-heads.

"In the most common and apparently original form, the flower-heads consist of florets which are all of the same tubular shape, and all contain both fully developed anthers and stigma, the divergence of the outer florets giving to the whole head a diameter of 20–30 mm. From this original form variation has gone on in two opposite directions, the final effects of this variation being, on the one side, very conspicuous male flower-heads of 50–55 mm. diameter; and on the other side less conspicuous female flower-heads of

---

[*] From an analysis of Ash blossoms, by Professor Church, *Journal of Botany*, 1877, p. 364.

[†] *Nature*, vol. xxv., p. 241.

30–35 mm. diameter. In both these extreme forms the outer row of florets possesses greatly enlarged radiating corollas which are sexually functionless, but useful in making the flower-mass more conspicuous. In the male flower-heads, anthers and pistils of the disk-florets are well-developed, but the style-branches never open so as to expose their stigmatic surfaces, and in their basal portion are grown together. In the female flower-heads, on the contrary, only the pistil of the disk-florets is fully developed, the anthers being pollenless, shrivelled, and brownish coloured.

"These two extreme forms are linked with the original one by a continuous series of gradations When in the original form variation begins in one direction, the outer row of florets gradually becomes longer and more radiating, and in the same degree their sexual organs diminish in size and become functionless, the anthers first aborting, and then the pistil. Finally, the barren ray-florets continuing to increase, the pistils of the disk-florets, too, become functionless, and the conspicuous male flower-head is accomplished.

"In the contrary variation some of the outer florets of the original form begin to diminish in size, while their anthers become brownish and pollenless, and this change step by step proceeds inwards and seizes a greater and greater number of disk-florets, until the whole flower-head is female, and reduced to a diameter of 15–18 mm. This state being reached, the corollas of the marginal flowers recommence to increase and become radiating, while at the same time their anthers disappear without leaving any trace, and their style-branches remain closed together."

Calendula officinalis furnishes another instance of complete change of sex, most probably caused by varying conditions of nutrition supplied by the soil. In the normal "single" form the disk florets are male, but with club-shaped stigmas. The two style arms, being fused together and strongly papillose, are only useful for thrusting out the pollen from the anther cylinder. In "double" forms the corollas all become ligulate, the stamens disappear altogether, and the style arms of the pistils assume the normal form characteristic of the ray florets. They now set seed, so that the entire capitulum is female, and forms fruit.*

Polygamous states often occur in trees growing apparently under the same conditions, and although we cannot doubt that they are due to different degrees of nutrition, yet they cannot be readily correlated to visible differences in the environment. Mr. Darwin thus describes the Ash: † "I examined fifteen trees growing in the same field; of these, eight produced male flowers alone, and in the autumn not a single seed; four produced only female flowers, which set an abundance of seeds, three were hermaphrodites, and two of them produced nearly as many seeds as the female trees, whilst the third produced none, so that it was in function a male. The separation of the sexes, however, is not complete in the Ash; for the female flowers include stamens, which drop off at an early period, and their anthers, which never open or dehisce, generally contain pulpy matter instead of pollen. On some female trees, however, I found a few anthers containing pollen-grains apparently sound On the male trees most of the flowers include pistils, but these likewise drop off at an early period, and the ovules, which ultimately abort, are very small compared with those in female flowers of the same age."

It may be added that the stamens are sometimes sub-

* I found no difference whatever between the plants raised from the larger seeds of the ray florets and the smaller ones of the disk florets. They all gave rise to the "single" form of capitulum.

† *Forms of Flowers*, p. 11.

petaloid forming staminodia—another hint that "contabescence" is closely akin to petalody of the andrœcium.

SEXUALITY AND HETEROGAMY.—Another source of diclinism may theoretically be attributed to protandry and protogyny carried to such a degree that the opposite sex is arrested altogether. Many plants have their flowers hovering about homogamy, some individuals being protandrous, others protogynous, according to locality, etc. Thus Saxifrages and species of *Ribes* are in this condition.

We know that as soon as a flower is fertilised, the corolla fades and mostly falls. This means that the nourishment is now directed into the pistil. In a protogynous flower the petals and stamens may be in a very undeveloped state, while the stigma is ready for pollination.* If it be fertilised it no longer requires other organs, and nourishment may be abstracted from the corolla and stamens, which therefore would tend to abort. Let this procedure become hereditary, and we get passages to female flowers. Moreover, the more female forms tend less to degeneracy, plant for plant, than the hermaphrodites, as Darwin showed with *Satureia*, and as is known to be the case with Strawberries in the United States, and again as is the case with the Ash, described above. Therefore female plants might be produced abundantly which would keep that form permanent.

Conversely, plants growing in the open with an increase of temperature, and readily seen and visited by insects, become strongly protandrous; consequently the pistil is at first delayed in development with a corresponding tendency to enfeeblement in comparison with the more purely female plants.

The results of crossing these conspicuous flowers—and

* See *e.g.* Müller's figures of *Saxifraga Seguieri* in different stages, *Fertilisation*, etc., p. 244.

the more conspicuous the more masculine is the flower, and the more attractive will it be—one with another, would not therefore be so advantageous as crossing the more female plants with the conspicuous. The former, too, produce relatively more offspring, and might tend to oust the others, and reproduce both the "more masculine" and the "more female" sorts. Intercrossing, therefore, coupled with environing conditions, may together bring about diœcism, as in Strawberries. As this reasoning is rather *deductive*, it must be only considered as a suggestion.

SEXUALITY AND HETEROSTYLISM.—This undoubtedly is another source of diclinism, as already alluded to. Mr. Darwin alludes * to *Coprosma* and *Mitchella* as indicating this fact. "Coprosma is diœcious, and in the male flowers the stamens are exserted, and in the female flowers the stigmas; so that, judging from the affinities of these genera, it seems probable that an ancient short-styled form, bearing long stamens with large anthers and large pollen-grains (as in the case of several Rubiaceous genera), has been converted into the male *Coprosma;* and that an ancient long-styled form, with short stamens, small anthers, and small pollen-grains, has been converted into the female form. According to Mr. Meehan,† *Mitchella repens* is diœcious in some districts: for he says that one form has small sessile anthers without a trace of pollen, the pistil being perfect; while in another form the stamens are perfect and the pistil rudimentary. *Mitchella*, therefore, would seem to be heterostyled in one district and diœcious in another," and this can scarcely be due to anything but environment.

* *Forms of Flowers*, etc., p. 285. See also above, p. 228.
† *Proc. Acad. of Sci. of Philadelphia*, July 28, 1868, p. 183. I do not gather from Mr. Meehan's account that he found any difference as to locality. Diœcism appears to be a constant character.

SEXUALITY AND THE ENVIRONMENT. 245

Summarizing the various influences of the environment as climatic—such as temperature and light, shade and obscurity, humidity and drought, as well as varieties of soil and degrees of nourishment, and possibly others—we soon see how careful one must be in attributing a result to any one or special cause alone. What we can do is, as it were, to pick out of them, as tolerably well-ascertained, conditions which seem to favour, say, the female as compared with the male organs or flowers—such as, e.g., a mean or *optimum* condition of vegetative energy, a relatively low temperature, no excess of nutriment, a due amount of light, humidity, etc.; or again, on the other hand, a relatively higher temperature, which favours and stimulates the staminal energies, the androecium being more keenly sensitive and more readily responsive to slight increments of temperature than is the gynoecium. The duration of the male elements being shorter than that of the female, they can come more quickly to maturity and perish earlier, as seen, for example, in the first flowering deciduous male catkins of *Castanea Americana* mentioned above. These, having been formed at the close of the preceding year (like many male flowers of the *Umbelliferæ* late in the season), may represent the expiring energy of the year's growth. They open first, as soon as a sufficient though slight increment of temperature occurs, but quickly fall off, quite useless, as no female flowers are open to be benefited by them.

Again, many, if not the majority of gynodioecious plants would seem to be produced by the first flowers opening before the temperature was sufficiently high to allow of the corolla and stamens to develop properly; and though many female flowers of the *Labiatæ* now blossom simultaneously with the hermaphrodite flowers of the same species; this may be, perhaps, accounted for by hereditary influences, as

Mr. Darwin showed that seeds of the female plants of Thyme yielded both female and hermaphrodite plants.

Although, therefore, we are unable to fathom all the mysteries of Nature's procedure, we can detect some of the lines upon which she works, and perceive how, in all cases, it is the environment—but sometimes one set of influences, sometimes another—which, being brought to bear upon the plant, the latter responds to it; and some form of what may be called "incipient diclinism" is the first result. If, then, these influences be kept up, hereditary conservatism comes into play, and such slight beginnings towards a separation of the sexes becomes fixed—only temporarily, however,—which constitute the first step, to be followed by others, till absolute and almost irrevocable diœcism is the final result.

Dr. M. T. Masters has collected several cases in which one or other of the sexes has been arrested, apparently in consequence of the nature of the soil and other conditions of the environment. I refer the reader to his "Teratology," as my object is not merely to enumerate all the instances known, but sufficient to establish the theory advanced,—that it is the environment that first influences the organism, which then responds to it; and that, secondly, all adaptive variations thus set up—provided the environment continue to exert its influences—can become fixed by heredity. The consequence is that they are ultimately recognized as constant and specific characters.

THE ORIGIN OF SEX.—If now the environment has been proved to exert potent effects upon the *development* of the sexual apparatus of flowers, there still remains the question how far is either sex or both present, or at least potential, in the embryo. Marked differences have resulted from sowing fresh or well-matured and older seeds of melons.

SEXUALITY AND THE ENVIRONMENT. 247

M. Arbaumont found that young seeds gave rise to plants of extraordinary vegetative vigour; moderately aged ones gave rise to corresponding moderately vigorous plants with both male and female flowers; while older seeds gave rise to still less vigorous plants, but which, when properly nourished, formed female buds.* M. F. Cazzuola † also found that melons raised from fresh seed bore a larger proportion of male flowers than female ; while older seed bore more female flowers: and this has been confirmed.

Another interesting result was obtained by M. Triewald, who grew twenty-one out of twenty-four melon seeds which were forty-one years old. The branches were very narrow, yet they produced early and plenty of good melons.‡ A cause of the differences of vigour in the plants raised from seeds of different age is, perhaps, connected with the fact that fresh melon seeds contain a neutral oil, which becomes more and more acid by keeping. This increased acidity coincides with a diminished germinative power; § and proportionately, therefore, less liable to run into excessive vegetative growth.

The next condition to be observed is that resulting from sowing seeds of diclinous plants thickly or thinly. Hoffman's experiments ‖ in this direction showed that 283 male

* *Bull. de la Soc. de Bot. de Fr.*, 1878, p. 111.
† *Bull. de Tuscan. Hort. Soc.*, 1877.
‡ *Gard. Chron.*, 1879, p. 470.
§ M. Ladureau in *Ann. Agronomiques*. Mr. Darwin also found that fresh seeds of *Iberis* grew at first more vigorously than others (*Cross and Self-fertilisation*, etc., p. 103).
‖ *Gard. Chron.*, 1879, p. 762; see also *Bot. Zeit.*, xliii., 1885, p. 145, seqq.; also *Jenaisch Zeitschr. f. Naturwiss*, xix. (1885), sup. ii., pp. 108–112. The following were the plants with which he experimented : *Lychnis diurna, L. vespertina, Valeriana dioica, Mercurialis annua, Rumex Acetosella, Spinacia oleracea*, and *Cannabis sativa*.

plants appeared, and 700 female, in the thickly sown plot, while only 76 males occurred when thinly sown. This has been paralleled in America, where Mr. Meehan, of Philadelphia, has noticed how *Ambrosia artemisiæfolia*, if growing vigorously, has a proportion of female flowers largely in excess of the males; but in fields where the grain has been cut, and this "Rag-weed" comes up in thick masses late in the season, the individual plants nearly starving each other, male flowers are very numerous, and some are wholly male. Prantl also observed that the crowded prothallia of Ferns gave rise to more antheridia, and scattered ones more pistillidia. Pfeffer, too, noticed the same fact with *Equisetum*.

. In these cases we seem to have results exactly the reverse of those of the melon seeds: but while in the latter the male flowers were accompanied by the precocious and excessive vegetative energy, the female were prevented from appearing at all; for it must be remembered that normally male flowers of melons appear before the females. In the case of thin sowing, the plants were in a natural and healthy condition: but when crowded they were starved, and the vital energy, being just enough to develop male flowers, proved insufficient for the female; and, conversely, when thinly sown, ".vitality" was not checked, and females were abundant.

The question arises, are all seeds potentially bisexual, and one sex rather than another determined either by an inherent vigorous constitution or by the conditions of the environment during germination and growth? or is there, so to say, a determination of sex, or at least a predisposition, at an earlier stage still? Dr. Hoffman, judging from his experiments, is inclined to the opinion that sex does not reside in the seed, but depends on conditions of germination. Mr. W. G. Smith arrived at the same conclusion, for he says in his *Remarks on some Diœcious Plants*,* "I think *seeds* themselves are probably not either male or female, but that *after influences* produce the sex; as in animals the sex is not developed in the early embryo life of the creature, nor till the embryo has attained a certain age."

On the other hand, F. Heyer thought sex was determined at an earlier period than the ripening of the seed.† Some differences which have been noticed in seedlings of Nutmegs seem to countenance this idea; thus Mr. Prestoe, in his report on the Trinidad garden,‡ says that "the leaf of the female seedling is most perfectly elliptical, with straighter primary veins. In the male plant it is broader towards the point than at the middle, *i.e.* obovate, and furnished with a point much longer than that of the female. The veins are also curved in towards the point much more roundly than in the latter."

An interesting experiment by Mr. I. Anderson-Henry, recorded in the *Gardener's Chronicle* of 1876, may be quoted. He says, "I raised a seedling *Begonia* having female flowers only. It resulted from an experiment I made on the seed-bearer by cutting off two of the three lobes which compose the stigma, and fertilising the remaining lobe. I repeated this experiment; and all of the progeny which have yet bloomed, consisting of four or five plants, have likewise all come with female flowers only." This seems to show that the female seedlings were due to *concentration of energy* to a limited number of seeds. On the other hand, a hybrid *Begonia*, "Adonis," raised by Mr. Veitch from a summer-flowering tuberous variety, "John Heal," crossed with a winter-flowering variety (itself obtained from *B. Socotrina* crossed by a dwarf-flowering tuberous variety), bore nothing

* *Journ. of Bot.*, 1864, p. 232 (note). † *Journ. Micr. Soc.*, 1884, 251.
‡ *Gard. Chron.*, 1884, p. 315.

but male flowers—presumably in consequence of some weakness of constitution due to hybridisation.

It would be quite foreign to my purpose to trace the origin of sexes throughout the vegetable kingdom, as I am solely concerned with that of flowers. But what appears to be pretty certain is that the absorption of the pollen-nucleus by the "egg-cell" involves a special form of nutrition, coupled with certain excitant effects. Union between nuclei occurs elsewhere; and as illustrative analogies, one recalls the fact of fusion being normal in the *Conjugatæ*, and among zoöspores, where no sexual differentiations are observable. Again, in the embryo-sac there occurs the union of two nuclei, one from each tetrad, their function being then apparently to form endosperm. As another case, Mr. Gilburt has described the union of the nuclei of cells constituting a "cell-group," which forms a wood-fibre after the absorption of the septa.*

Of course one of the most essential properties of the pollen-nucleus is to transmit to the offspring characteristics of the male parent: but even this is paralleled in the vegetative system; for an engrafted scion can transfer its peculiarities to the stock, as has occurred with *Cytisus Adami*, variegated Abutilons, etc.

If, however, we ask what are the actual differences which exist between the male and female energies, and how they have arisen, we at once find that we are completely baffled, and that all speculations are at present futile.

* *Morph. of Veg. Tiss.*, Journ. Roy. Micr. Soc., 1879, p. 806 (note). Schacht observed a similar origin of liber-fibres in the Papaw, each of which was originally composed of three or four cells, but the septa become absorbed; their original positions being only indicated by clusters of pores on the walls (*Les Laticif. du Carica Papaya*, Ann. des Sci. Nat., 4 sér., viii., pl. 8, figs. 9, 10). Treub, on the other hand, discovered the laticiferous vessels and liber-fibres of the Nettle, etc., to have arisen by repeated division of the nucleus, the partitions not having been formed at all (*Arch. Neerl. des Sci. Exac. et Nat.*, tom. xv., 1880, p. 39).

CHAPTER XXVI.

DEGENERACY OF FLOWERS.

INCONSPICUOUS AND CLEISTOGAMOUS * FLOWERS.—Degeneracy in plants is as of frequent occurrence as in animals; and just as it implies no pathological or anything of a constitutionally injurious character in them, so, it must be distinctly borne in mind, does it imply nothing of the sort in plants. The word means "down from the genus;" like "degradation," it is only a "step downwards." It implies retrogressive or at least arrested conditions; but a degraded flower often acquires new features, qualifying it for securing self-fertilisation with a far greater certainty than was the case with its more conspicuously flowering ancestors.

There are several causes which can bring about degradations in the various organs of plants, such as growth in water, subterranean habits, parasitic and saprophytic states, freedom from strains, compensation, etc. Though it would be interesting to trace out the cause and effect in each case, I must content myself with flowers, and particularly the essential organs.

There are two principal causes which may be styled the rationale of degradation in flowers. The first is compensation, when the vegetative system is in too great activity to

* *Cleistogamous*, "a closed union," *i.e.* when flowers are self-fertilising without opening.

allow of the proper amount of nutrition being at the service of the flowering process. This is so well known that I need not dwell upon it now. The second is the cessation of insect fertilisation. The effect of fertilisation operates in two directions. On the one hand, if it be the result of intercrossing by insect agency, it stimulates the flowers till they become thoroughly adapted to their visitors, and highly differentiated in certain ways in consequence, but more especially as regards the perianth and stamens; while, in many cases, some degree of degradation occurs simultaneously in the pistil. Conversely, self-fertilisation and anemophily, consequent upon the neglect of insects, are accompanied by corresponding degradations in the perianth, stamens, and pollen, correlated with a regained ascendancy in the powers of reproduction. The limits of degradation, with an increase of fertility, are seen in many cleistogamous flowers.

In tracing the progress of degeneracy from a species with large flowers to one with inconspicuous blossoms, I do not mean to imply that we can actually witness the process in activity· but we can see this represented, as it were, in many a series of what we call species of a genus; but which we might call transitional forms of one kind. It is only because we cannot trace the actual process going on that we regard them morphologically as distinct species. Thus, if a verifiable demonstration be unattainable, it is a "moral conviction," not only that *Geranium pratense* is as much and obviously adapted to insect agency as *G. pusillum* is to fertilise itself, but that the latter species has been derived from the former or from some kindred plant, through some such transitional forms as *G. pyrenaicum* and *G. molle*.

This process of degradation from insect to self-fertilising conditions, not only affects the size of all parts of the flower, but the entire plant. Mr. Darwin showed how the stimulating effect of crossing generally increased the heights and weights and, for a time, the fertility of the plants experimented upon. Conversely, self-fertilised species are altogether smaller than their allied intercrossing species. Thus *Stellaria Holostea* may be compared with *S. medium, Cerastium arvense* with *C. tetrandrum* and *C. glomeratum, Cardamine pratensis* with *C. hirsuta, Polygonum amphibium* with *P. aviculare,* etc. Besides being thus dwarfed, self-fertilising plants are mostly annuals. But while conspicuous flowering plants blossom during a limited period in summer only, their smaller, less conspicuous, and regularly self-fertilising allies may, and often do, flower and set seed all the year round.

In my essay on "The Self-fertilisation of Plants," * I drew up the following list of peculiarities of habitually self-fertilising plants, all of which indicate points of degeneration or arrest.

1. The inconspicuousness of the flowers, even when fully expanded.

2. The calyx and corolla are often only partially expanded, or not at all.

3. The white or pale colours of the corollas; while specially coloured streaks, specks, "guides," and "path-

---

* I must refer the reader to the above essay for a full discussion of this subject. The evidence there given proves conclusively that self-fertilising and anemophilous plants are in every way the most widely dispersed of flowering plants, and best fitted to maintain themselves in the struggle for life. I will add here that Mr. H. O. Forbes came independently to a similar conclusion when studying cleistogamy in orchids; and remarks, at the close of his paper (*Journ. Lin. Soc.*, vol. xxi., Bot., p. 548), "The observations above given would seem, therefore, to support the Rev. G. Henslow's conclusions so ably given in his 'Memoir on the Self-fertilisation of Plants,' already published in the Transactions of the Linnean Society. My absence abroad prevented my seeing this paper till quite recently, and after I had completed these notes."

finders" peculiar to intercrossed flowers are more or less reduced, if not absent.

4. The partial or total arrest of the corolla.

5. The mature stamens of the expanded flower retain in many cases the incurved, *i.e.* an arrested position, which they had in bud; the anthers thus remain in contact with the stigmas.

6. The stamens are often reduced in size and number, and the pollen in quantity.

7. The pollen tubes may often be seen to be penetrating the stigmas, either from grains still within the anther-cells, or evidently derived from those of the same flower.

8. The styles are shortened, and the stigmas are situated appropriately for direct pollination from the anthers of the same flower.

9. The partial arrest of the corolla and stamens in their rates of development, allows the pistil to mature with comparative rapidity.

10. The consequent early maturation of the stigma, so as to be ready before or simultaneously with the dehiscence of the anthers.

11. Little or no scent.

12. Decrease in size or total absence of honey glands, with corresponding little or no secretion of honey.*

Notwithstanding these various indications of degradation, such flowers are often correlated with special alterations which secure self-fertilisation without a chance of failure —a precariousness which almost always exists in flowers adapted to insects. Thus—contrary to the old but erroneous

---

* Müller, in his "General Retrospect" (*Fertilisation*, etc., p. 591), also gives a number of modifications, mostly referred to in the text above, of what he describes as "the countless ways in which plants revert to self-fertilisation in default of sufficient insect visitors."

DEGENERACY OF FLOWERS.

dictum that, whether flowers were pendulous or erect, the stigma was always *below* the anthers, so that pollen could *fall* upon it—the anthers are always *closely applied* to the stigmas, as may be seen in Chickweed (Fig. 52), and small-

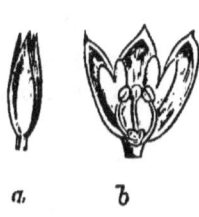

Fig. 52.—Flower-bud, closed and expanded, of *Stellaria media*, showing petals reduced in size; stamens, three only; anthers closely adpressed on stigmas.

Fig. 53.—Stamens and stigmas of *Epilobium montanum*, the bud scarcely open, while anthers are closely applied to the stigmas.

flowered Willow Herbs (Fig. 53), and especially in cleistogamous flowers (Figs. 56–59, pp. 258–261).

The structure of the anthers and stigmas is often greatly altered in form, besides being merely reduced in size.

As an illustration of the above remarks, the genus *Viola* is interesting as furnishing two "forms" of the same species, *V. tricolor*, or Pansy, the one being adapted to insects, the other to self-fertilisation; while other species, such as *V. odorata*, the Violet, bear cleistogamous buds on the same plant as the ordinary violet blossom.

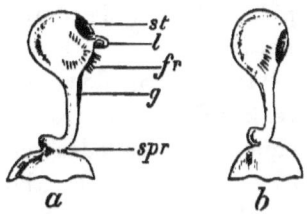

Fig. 54.—Styles and stigmas of the two forms of Pansy · *a*, that of the larger and intercrossing; *b*, that of the self-fertilising form.

The dimorphic flowers of *Viola tricolor* were first noticed by Müller, who described them as follows: * "In the large-flowered form, the stigmatic cavity (Fig. 54, *a*, *st*) lies some-

* *Nature*, Nov. 20, 1873, p. 45.

what more towards the top of the skull-like end of the style than in the small-flowered one (*b*). When the skull-like knob in the two forms is pressed against the lower petal, in the large-flowered form the opening of the stigmatic cavity is directed outwards, so that the pollen-grains which have fallen out of the anther-cone can never spontaneously fall into the stigmatic cavity, and must be carried there by insects; whereas in the small-flowered form the opening of the stigmatic cavity is directed inwards, so that pollen-grains falling out of the anther-cone spontaneously, fall directly into the stigmatic cavity.

"In the large-flowered form, the opening of the stigmatic cavity (*st*) bears, on its lower side, a labiate appendage (*l*) provided with stigmatic papillæ, so that a proboscis inserted into the flower when charged with pollen from a previously visited flower, rubs off this pollen on to the stigmatic lip, thus regularly effecting cross-fertilisation; whereas, when withdrawn out of the flower, charged with pollen, the proboscis presses the lip (*l*) against the stigmatic opening (*st*), thus preventing self-fertilisation. This nice adaptation to those visitors provided with a long proboscis (Lepidoptera, Apidæ, Rhingia) is completely wanting in the small-flowered form (*b*).

"In the large-flowered form, there is a black wedge-shaped streak (*g*) on the front of the style, to which Mr. A. W. Bennett first called attention, and which he has interpreted as a guide-mark for those visitors which are diminutive enough to crawl entirely into the flower. This streak is also wanting in the small-flowered form.

"In the large-flowered form, pollen-grains do not spontaneously fall out of the anther-cone before the flower has been fully developed for several days; whereas, in the small-flowered form, in bb far the majority of cases, a great number

DEGENERACY OF FLOWERS.

of pollen-grains fall spontaneously out of the anther-cone into the stigmatic cavity and there develop long pollen-tubes, even before the opening of the flower, in much rarer cases a short time after it has opened.

"When the visits of insects are prevented by a fine net, the flowers of the small-flowered form wither two or three days after opening, every one setting a vigorous seed-capsule; those of the large-flowered form remain in full freshness more than two or three weeks, at length withering without having set any capsule; when fertilised they, too, wither also after two or three days."

I have met with several variations in minor details of structure in the smaller-flowered kind. Thus in some the stigmatic lip, probably representing one of the three stigmas, formed a globular knob protruding from the orifice, as shown in Fig. 55, *a*, *b*. In another, it protruded like a tongue, *c*. The lateral fringes,* which help to keep the pollen back from reaching the stigmatic chamber in the larger flowers, are more or less retained in these; as is also the bent-base to the style which forms the spring,* which keeps the globular head in a downward position.

Fig. 55.—Styles and stigmas of self-fertilising forms of Pansy. (For description, see text.)

The accompanying figures will illustrate the cleistogamous flower-buds of Violets. They are very minute, about one-eighth of an inch in length (Fig. 56, *f*). The petals are reduced to linear and pointed structures, green or purplish green, (*a*); or they may be altogether wanting. The spur alone of the larger petal is sometimes present in strong-

* For the theoretical origin of "fringes" and "springs," see Chap. XV., p. 133, and Chap. XIII., p. 123, respectively.

growing garden plants (*b*). The stamens are five or less in number, having spoon-shaped connectives, and not pointed as in the normal form, bearing very minute oval anther-cells at the base (*c, g*) * Small bundles of pollen-tubes may be traced from the anthers into the stigma (*g*). The pistil has a short curved style, and truncated stigma (*d*) concealed beneath the anthers which lie imbricated over the top of the pistil. The anthers are usually devoid of appendages, though they are sometimes present, like the spur; though

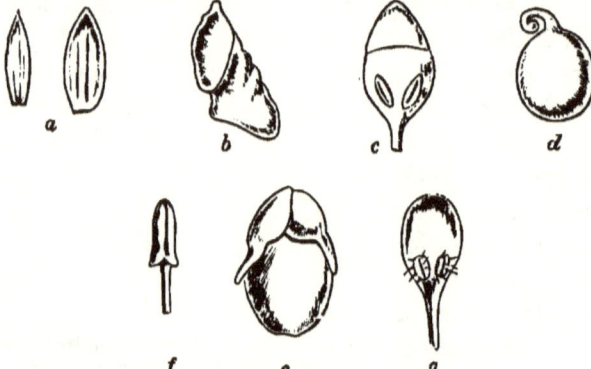

Fig. 56.—Cleistogamous Violets. (For description, see text.)

both organs are now useless. As the ovary swells it raises the stamens up with it (*e*). The capsules of the violet, Mr. Darwin observes, bury themselves in the soil, if it be loose enough, and there ripen; but they certainly are very, if not more frequently not buried at all, but only concealed beneath the foliage.

As another interesting case of a plant showing transitional conditions may be mentioned *Scrophularia arguta*, Ait.† "The two lowermost opposite and axillary branches bend backwards and penetrate the soil. The next pair do

* (*c*) *V. odorata;* (*g*) *V. canina.*   † *Bull. Soc. Bot. de Fr.*, iii., p. 569.

the same, but do not always reach the ground, or else penetrate it very slightly. They all bear fertile flowers. The lowest are apetalous, if completely hypogean [and presumably cleistogamous]. Those which just reach the soil have a corolla of four lobes nearly equal, and resemble the corolla of *Veronica*. A little higher up, the irregularity of the bilabiate character of *Scrophularia* is pronounced."

The preceding quotation is interesting, first in showing how the subterranean cleistogamous form is derived from the conspicuous flower, and also supplies a hint as to the origin of *Veronica*, in that it is a 4-merous degradation from a primitive 5-merous genus, which is lost or unrecognizable now, unless it be some member of the subgenus *Pygmæa*, which has five parts to the corolla.*

As an illustration where geographical conditions favour the development of autogamous forms of flowers, the following passage may be quoted :—

"Herr C. A. M. Lindman has examined the very rich flora of the Dovrefjeld in reference to the arrangements for fertilisation. He finds a distinct tendency to a deeper colour in the flowers than is displayed by the same species in the lowlands, red and blue predominating. The great length of daylight appears to increase the size both of leaves and of flowers, though in some species, on the other hand, the flowers are diminutive in consequence of the low temperature. Crowded masses of small flowers are very common. The number of scented species is comparatively small, though the fragrance is sometimes powerful. The scarcity of insects necessitates that there should almost always be a provision for possible self-fertilisation; and many species, elsewhere heterogamous, are here homogamous. Notwith-

* For Müller's theory of the origin of *Veronica*, see *Fertilisation*, etc., p. 465.

standing the cold and wet summer (1886), the plants observed almost invariably bore fruit."*

As an example of pure cleistogamy I will take *Oxalis Acetosella*, as having special peculiarities. Mr. Darwin alludes to M. Michalet's description of the cleistogamous flowers of this species,† and adds some observations of his own.‡ He quotes an observation of Michalet's, that the five shorter stamens are sometimes quite aborted. This fact, which I have also observed (Fig. 57, *d*), is quite in keeping with the common process of the reduction of the number or parts of stamens in self-fertilising flowers.

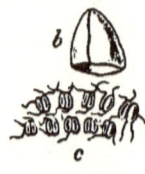

Fig. 57.—Cleistogamous flower-buds of *Oxalis Acetosella*. (For description, see text.)

He also adds this interesting observation: "In one case the tubes, which ended in excessively fine points, were seen by me stretching upwards from the lower anthers towards the stigmas, which they had not as yet reached. My plants grew in pots, and long after the perfect flowers had withered they produced not only cleistogamic, but a few minute open flowers, which were in an intermediate condition between the two kinds." This last remark is quite in accordance with the true origin of these flowers, that they are in all cases degradations from the conspicuous forms normally characteristic of the species which produce them.

Fig. 57, *a*, clearly shows that in *Oxalis Acetosella* the cleistogamous state is simply a flower-bud which has become *adapted* to self-fertilisation; and the intermediate conditions alluded to by Mr. Darwin I should suspect were analogous to

* *Journ. Roy. Micr. Soc.*, 1887, p. 615, and note. See below, pp. 270, 271.
† *Bull. Soc. Bot. de Fr.*, vii. (1860), p. 465.
‡ *Forms of Flowers*, p. 321.

DEGENERACY OF FLOWERS.

the permanent forms of the flowers of *O. corniculata*, which I at first inferred, from the wide distribution of this species, must be habitually self-fertilising. From Fig. 57, a, it will be seen that the corolla just protrudes from the closed sepals, and always remains as a "cap," b. Of the ten anthers, five are often abortive or wanting, d; the fertile anthers are placed over the very short stigmas, and are bound together by fine threads. These appear to play some part, but the nature of their function is obscure, c.

Impatiens fulva and *I. Noli-me-tangere* have also cleistogamous flowers. Fig. 58, a, represents a bud, and b two metamorphosed stamens.

Lamium amplexicaule will furnish another example of cleistogamy. This genus has usually flowers highly differentiated, and adapted to insect fertilisation. That the cleistogamous flowers of this, as of all other species, are degraded forms of the normal kind is obvious from the presence of the "lip," as well as by there being four and didynamous stamens. The style elongates very much, and under the pressure of the closed summit of the corolla becomes bent, so that the stigmas lie between the anther-cells, and thus readily become fertilised. Fig. 59, a, represents a flower-bud; b, the corolla in section; and c, the pistil removed. This condition of cleistogamy is found in the earlier-flowering plants, so that it is probably a mere result of check through a colder temperature.

Fig. 58.—a, Cleistogamous flower-bud of *Impatiens fulva*; b, stamens (after Bennett).

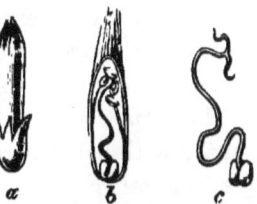

Fig. 59.—a, Cleistogamous flower-bud of *Lamium amplexicaule*; b, vertical section of same; c, pistil.

Salvia clandestina may be compared with the last described, as it is a self-fertilising form of, perhaps, *S. pra-*

tensis. Fig. 60,* *a*, represents a corolla, which is very small, but open; *b* represents the two fertile stamens; the anther-lobes instead of being horizontal are erect, and face each other. The stigmas curl back between them, and are remarkably long, *c*.

Fig. 60.—*Salvia clandestina:* a, corolla; b, anthers; c, style and stigmas.

THE ORIGIN OF CLEISTOGAMY.—We are now in a position to trace the causes of cleistogamy. Cleistogamous flowers nearly always occur on plants otherwise, or at least their allied species are, adapted for intercrossing, and include four genera of anemophilous plants. The first cause or influence is the arrest of the reproductive energy in the conspicuous flowers, which often set no seed at all.

Whatever the primary cause of that may be, a very common result in perennials is to increase the power of vegetative methods of multiplication, as in the case of many bulbous and tuberous plants.

This, however, is not a special feature of the plants which bear cleistogamous flowers. It would seem, therefore, that the reproductive energy being checked in one form of flower, it, so to say, breaks out in another. But there are several influences at work, and a very obvious one is temperature; for the same species may behave very differently in one country with a high mean annual temperature, from what it does in another with a lower one. Thus, *Viola odorata* does not produce cleistogamous flowers in one part of Liguria, where the conspicuous flowers are perfectly fertile; while they are mostly barren in England. On the other hand, cleistogamous flowers are produced by Violets near Turin,

* From a specimen growing at Kew. It is cleistogamous at Halle (see below, p. 263).

and abundantly in all parts England. *Viola nana* bears normal flowers in its native home in India, but only cleistogamous ones in England. *Viola palustris* bears only the larger flowers near Paris, which are perfectly fertile, but when it grows on mountains it bears cleistogamous flowers. Similarly *Impatiens fulva* bears both kinds of flowers in England, but the larger are usually barren. After midsummer, in its native home in the United States, these flowers will produce capsules. *Salvia clandestina*, when transplanted from Africa to Halle, bore only cleistogamous flowers for five years, according to Ascherson, who considered the plant to afford an example of continuous self-fertilisation. He, however, afterwards observed ordinary open flowers. It is a species particularly common on the Continent.

Again, plants vary according to the season. Thus Mr. Darwin found that *Vandellia nummularifolia* bore no perfect flowers in one season; so, too, *Ononis columnæ* bore none in 1867, yet it had both kinds in 1868.

The time of the year also influences the production of cleistogamous flowers. Thus *Ononis minutissima*, *O. parviflora*, and *O. columnæ*, according to Mr. Bentham, produce them early in the spring. *Godetia Cavanillesii* and *Lamium amplexicaule* do the same; while some bear a fresh crop in the autumn, as *O. columnæ*.

Two cases are mentioned by Mr. Darwin in which the period is the reverse of the above.

Viola Roxburghiana bore abundance of cleistogamous, but no perfect flowers, in Mr. Darwin's hothouse; and it bears the perfect flowers in India "only during the cold season, and these are quite fertile. During the hot, and more especially during the rainy season, it bears an abundance of cleistogamous flowers."*

* *Forms*, etc., p. 320.

The other example is *Ruellia tuberosa*, of which Mr. Darwin remarks, "It produces both open and cleistogamous flowers; the latter yield from 18 to 24, whilst the former only from 8 to 10 seeds: these two kinds of flowers are produced simultaneously, whereas in several other members of the family the cleistogamous ones appear only during the hot season." From this one would infer that an excess of heat may be a cause of cleistogamy, just as too low a temperature appears to bring it about.

I think it probable that other influences than temperature may be brought to bear upon a plant; which, indeed, we may see in our own Violets. The larger flowers of this species are *not* produced in the hottest time of the year, while the cleistogamous buds are only borne in the summer. On the other hand, the foliage is only developed fully, contemporaneously with the dwarfing of the floral organs.

Again, a poor soil has been noticed as associated with cleistogamy by Torrey and Gray, in the case of North American species of *Helianthemum*.

Temperature, however, seems to be the most important agent; thus, while the climate of South Italy can develop the perfect flowers and *render them fertile*, there cleistogamy is suppressed; here, in England, the climate is seemingly not sufficiently warm to do so, and the cleistogamous buds appear in compensation. The vegetative energy, however, comes to the fore during the summer, and perfect flowers are not produced simultaneously with it; so that it is not until the vegetative period has ceased, and the materials are remade for their development, that larger flowers are again borne later in the year, as in November, as well as in the following spring.

With regard to the anemophilous genera, Mr. Darwin mentions *Hordeum, Cryptostachys, Leersia oryzoides*, and *Juncus bufonius* in Russia.

Now, the three genera of Grasses here mentioned are characteristic of warmer regions, and even tropical, *Leersia oryzoides* being the sole species of that genus which reaches Europe, where it becomes cleistogamous. Therefore climatal conditions may, with some reasonable presumption, be suggested as the immediate cause in these cases. With regard to *Hordeum murinum*, which is, perhaps, almost habitually cleistogamous in this country, it may be an hereditary result issuing from a similar cause. This may also apply to *Viola*; for not only are some species tropical, but all the genera most nearly allied to *Viola* are tropical also. This is analogous to what I have suggested as the origin of gynodiœcism in *Labiatæ*, which it may be noticed has at least two genera with cleistogamous flowers in this country or Europe. *Juncus bufonius*, according to Batalin, is exclusively cleistogamous in Russia, hence the same cause suggests itself for this species; for, according to Ascherson, at Halle it has ordinary open, lateral, hexandrous flowers in addition to terminal cleistogamous triandrous ones.* This seems to show that lessened vigour has also a hand in the process in this case: the mean temperature of Halle is probably higher; if so, it may cause the plant to bear the open flowers there.

From the above-mentioned facts, it will be seen that there may be more than one cause to account for cleistogamy. Hence, it must be regarded as an *inevitable result* whenever those influences are brought to bear upon the plant which are capable of producing it; and there is every reason to believe that whatever effects are produced in plants by external stimuli, if the latter be permanently kept up they will become hereditary, and then will be recognized by systematists as specific or generic characters.

ANEMOPHILOUS, OR WIND-FERTILISED PLANTS.—The general

* Müller, *l.c.*, p. 561.

characters prevailing in this group consist of elongated papillose or plumose stigmas, or else they spread out into laminæ (*Euphorbia*). The filaments are usually slender and movable, with versatile anthers, bearing incoherent and often smooth pollen-grains. In some cases the filaments are elastic, and project the pollen outwards; or the whole flower may oscillate on a slender pedicel or peduncle, as the catkins of the *Amentiferæ*, the flowers of *Rumex*, etc. Long, slender filaments are seen in Grasses, Sedges, Rushes, Hemp and Hop, Plantains, *Littorella*, and *Poterium*. Nettles and their allies are remarkable for their elastic filaments, which materially aid in the dispersal of the pollen.

On the other hand, Palms, Bulrushes, etc., have more or less rigidly fixed flowers and floral organs.

There is little doubt but that all wind-fertilised angiosperms are degradations from insect-fertilised flowers. This is obviously so when many of the allies of an anemophilous genus or species are constructed for insects. Thus, Müller says that *Thalictrum minus** is anemophilous, while *T. flavum* is visited by several species of insects. *Poterium Sanguisorba* is anemophilous; and *Sanguisorba officinalis* presumably *was* so formerly, but has reacquired an entomophilous habit; the whole tribe *Poterieæ* being, in fact, a degraded group which has descended from *Potentilleæ*. Plantains retain their corolla, but in a degraded form. *Junceæ* are degraded Lilies; while *Cyperaceæ* and *Gramineæ* among monocotyledons may be ranked with *Amentiferæ* among dicotyledons, as representing orders which have retrograded very far from the entomophilous forms from which they were possibly and probably descended.

* I do not know on what reason; for the stigmas are not characteristic of such flowers. On *à priori* grounds I should have inferred its being self-fertilising, as the anthers completely conceal the few and small carpels.

What, then, have been the causes which have given rise to the features generally characteristic of anemophilous flowers? In the first place, it must be remembered that such are far from absolute. Smooth and easily scattered pollen,* Müller remarks, is the only positive character common to these plants. Mr. C. F. White, F.L.S., however, tells me that from his researches he very much distrusts the division so generally accepted between wind- and insect-borne pollens. It is his opinion that there is no pollen-grain so smooth but that the hairs on the limbs of a bee or fly can hold it. Moreover, no pollen, however massed together, can possibly be heavier than, say, a thistle seed and its down attached, which the wind can carry with perfect facility; so that to draw any distinction on that score seems to me to be very far-fetched.† With respect to the pollen of Grasses, Mr. White observes that it is perhaps forgotten that, although smooth in water, when dry they are notably wrinkled into sharply angled and irregular shapes.

Mr. Edgeworth ‡ has figured many forms of pollen of anemophilous genera, several of which show no signs of smoothness or rotundity, such as *Alopecurus pratensis*, *Carex arenaria*, and *C. panica*, which, like *Juncus effusus*, is oblong, with sharp edges, all of which are at right angles or nearly so. Again, *Typha latifolia* and *Cupressus* have octahedral pollen; *Areca Baueri*, *Ceratozamia*, *Rheum*, *Mercurialis*, Oak, etc., have more or less sharply pointed spindle-shaped grains.

* See Mr. A. W. Bennett's paper, *On the Form of Pollen-grains in Reference to the Fertilisation of Flowers*, Brit. Assoc. Rep., 1874.

† I would here allude to another à *priori* assumption. It has been thought that the two pouches on the pollen of the Fir aid it in transportation; but unless they were filled with some gas lighter than air they only increase the weight of the grain.

‡ *Pollen*, by Mr. M. Pakenham Edgeworth, F.L.S., 1877.

In *Corylus, Alnus,* and *Plantago media,* they are polygonal, while Beech has them deeply three-grooved, etc.

Mr. Edgeworth, in fact, states that the different kinds of pollen of anemophilous plants " are by no means all globular, as Mr. Bennett asserts."

He notices, however, that " the grasses and *Cyperaceæ,* and perhaps the *Plantagineæ* are without the sticky nature of the outer coat, which obtains through all other pollen grains."

With regard to the versatile condition of the anthers in grasses, and their consequent facility of oscillating on a point, this feature seems to be only the result of the extremely slender filament due to degradation; * and not quite the same thing as the antero-posterior oscillation which the action of bees has set up in the connectives of *Salvia,* species of *Calceolaria,* and *Curcuma Zerumbet.*† Remembering how the rigidity of the filaments of intercrossing flowers is correlated to the retention of some well-defined positions for the anthers, so that insects can be struck by them accurately, and be again struck on the same spot by the stigmas of other flowers, we see that when the stimulus due to intercrossing has been long withheld, the filaments have become slender, easily waved about by the wind, and versatility of the

* *Plantago media,* which is visited, has motionless anthers; but in the anemophilous species of Plantain they are versatile.

† Mr. H. O. Forbes has described and figured a very analogous case in this species of *Curcuma* of Sumatra. The two anthers project forwards in contact, they are provided with terminal processes like horns. The style passes between them. When a bee enters the flower it depresses these horns with its head, and so forces the anthers downwards on to its thorax. The anthers bring the style and stigma down also. In a similar way do some species of *Salvia* cause the style to be brought down from the hood (*A Naturalist's Wanderings in the Eastern Archipelago,* p. 247).

anthers has followed. Those wind-fertilised plants with stiff filaments have presumably not yet degraded to a similar state.

With regard to the pistil, since of heterostyled plants the stigmatic papillæ are larger and longer in the long-styled forms, we seem to get a hint as to the origin of the papillose and plumose characters of many wind-fertilised plants; in that such may be due to compensatory processes on the loss of the corolla, honey-secreting organs, etc., which have thus favoured the development of the pistil generally, such developments becoming emphasized in certain directions.

Protogyny or homogamy generally accompany anemophily.* Thus Müller mentions *Thalictrum minus*, *Plantago*, *Luzula*, *Callitriche*, *Myriophyllum*, and many Grasses as being protogynous; and a common characteristic feature of such flowers is frequently noticed by Müller, viz., that they have all "long-lived stigmas." This seems clearly to point to a relatively increased amount of vigour in the development of that organ in protogynous flowers; which becomes especially noticeable in their enhanced size, as seen in most anemophilous flowers. *Poterium* he regards as homogamous, as well as Rye and Wheat. These conditions all agree with the total suppression of the corolla, and may be regarded as signs of degradation: and I have elsewhere shown, when treating of emergence and development of the floral organs, how a compensatory process accompanies the formation of the corolla and stamens on the one hand, and of the pistil on the other; so that when the former tend towards degradation, the pistil gains the ascendancy, and matures earlier.

* *Artemisia vulgaris* seems to be protandrous. The style arms are provided with papillose rosettes in the central florets, but are very elongated, and terminate in points in the circumferential florets. In no case could I detect pollen-tubes in unopened florets, though the grains were shed.

Hence, to find its stigmas enlarging under anemophily is all in keeping with the above facts.

THE ORIGIN OF ANEMOPHILY.—With regard to the origin of anemophilous flowers, there is every reason to believe them to be due to the neglect or absence of insects: that as these have brought about brilliant colours or other kinds of conspicuousness, so their absence has allowed flowers to degenerate and become inconspicuous, the result being either self-fertilisation or anemophily. As two examples of districts which illustrate this fact, are the Galapagos Islands, visited by Mr. Darwin, and Greenland, the flora of which is described by M. Warming.

The former observer, on landing, thought that there were few or no flowers, but, on stricter search, discovered many to be inconspicuous. A specimen before me of *Solanum nigrum*, which he brought from those islands, has flowers much smaller than our own native plant, and illustrates the wide dispersion of self-fertilising plants. M. Warming found Greenland, like the Galapagos Islands, to be poor in insects, and "the flowers display a corresponding increased tendency to autogamy. One hundred and thirty-eight species of anemophilous plants are also named by him, exclusive of Willows. The flowers appear to decrease in size with the increase of latitude; and the brilliancy of colour certainly does not become greater." *

This last observation does not agree with M. Flahault's observations; † and possibly M. Warming is here intimating a wrong cause of degeneracy, which I should incline to regard as the absence of insect stimulation, with the consequent tendency to inconspicuousness, anemophily, and autogamy.

* *Overs. K. Danske Vidensk. Selsk.*, 1886, p. xxv. (quoted from *Journ. Roy. Micr. Soc.*, 1887, p. 433). See also above, pp. 177 and 259.

† *Ann. des Sci. Nat.*, 6 sér., t. vii. (1877), et t. ix. (1879).

Where, however, insects are abundant, whether in high latitudes or greater altitudes, as in the Alps, there two causes will be at work to enhance the brightness of flowers; viz. insect stimulation and prolonged sunlight. For Sachs has shown that the ultra-violet and invisible rays are specially efficacious in the development of flowers; and as the foliage grows more vigorously with prolonged light so it is presumable that the flower-forming substances will be more abundant as well.*

The genus *Plantago*, like *Thalictrum minus*, *Poterium*, and others, well illustrates the change from an entomophilous to the anemophilous state. *P. lanceolata* has polymorphic flowers, and is visited by pollen-seeking insects, so that it can be fertilised either by insects or the wind. *P. media* illustrates transitions in point of structure, as the filaments are pink, the anthers motionless, and the pollen-grains aggregated, and it is regularly visited by *Bombus terrestris* (Delpino). On the other hand, the slender filaments, versatile anthers, powdery pollen, and elongated protogynous style are features of other species indicating anemophily; while the presence of a degraded corolla shows its ancestors to have been entomophilous. *P. media* therefore illustrates, not a primitive antomophilous condition, but a return to it; just as is the case with *Sanguisorba officinalis* and *Salix Caprea;* but these show no capacity of restoring the corolla, the attractive features having to be borne by the calyx, which is purplish in *Sanguisorba*, by the pink filaments of *Plantago*, and by the yellow anthers in the Sallow Willow. *Plantago alpina* is self-fertilising, as the stigma does not wither until after maturing the anthers.

If we may speculate as to why some degraded flowers

* See *La Végétation du Globe*, par Grisebach, t. i., p. 155 (trad. fran. de Tchihatchef).

have become regularly autogamous, while others are now anemophilous, it may be due to the fact that, if a flower has been entomophilous and even strongly protandrous, the first stage of degradation is to bring the essential organs to a homogamous state. If they stop there, and become autogamous as well, which is the usual result, then the flower will remain persistently self-fertilising, as, *e.g.*, Shepherd's-purse, Chickweed, Knot-grass, etc.

If, however, the flower had been protogynous, such as early-flowering Hellebores, *Prunus communis* or some Alpine species, with "long-lived stigmas," then this protogyny, associated with other degradations of the corolla, etc., which only tend to increase it, has ended with anemophily.

In the first case the androecium of protandrous flowers has come down from its previous highly differentiated state, so as to be homogamous with the stigmas. From the other or protogynous condition, the gynoecium has not been brought back again so as to be homogamous with the anthers and pollen, but, on the contrary, it may have become even further differentiated, and so has now no fertiliser to depend upon except the wind.

CHAPTER XXVII.

DEGENERACY OF FLOWERS (*continued*).

DEGENERACY OF THE ANDRŒCIUM.—The number of stamens may decrease, as well as the quantity of pollen; while the form of the anthers may change and the character of the pollen may alter; and lastly, the position of the stamens may not be the same as in intercrossing flowers,—all these forms of degradation being so many adaptations or adjustments for self-fertilisation. They are well seen in Violets and the Wood-sorrel.

As examples, in *Stellaria Holostea* there are ten stamens, in *S. media* only three; and in cleistogamous Violets they vary from five to three or two. In the latter, the anthers become spoon-shaped with a rounded connective and much reduced anther cells; in the cleistogamous flowers of *Oxalis Acetosella* the pollen is almost deliquescent. Lastly, in all flowers especially adapted for self-fertilisation the anthers are in contact with the stigmas in consequence of their arrest in growth.

It must be noted here that this degeneracy in the stamens in no way impairs their functional value. The fact is that a very small amount of pollen is really quite sufficient for fertilising a considerable number of ovules.

For convenience I call it degeneracy, but another view would be to regard it as the *conservation of energy*, instead of wasting it in the production of a great deal more pollen than is usually required.

An interesting experiment of Mr Darwin's proves this. He placed a very small mass of pollen-grains on one side of the large stigma of *Ipomœa purpurea*, and a great mass of pollen over the whole surface of the stigmas of other flowers, and the result was that the flowers fertilised with little pollen yielded rather more capsules and seeds than did those fertilised with an excess.* That normally intercrossing flowers produce a great superfluity of pollen is well known. Thus Kölreuter found that sixty grains were necessary to fertilise all the ovules of a flower of *Hibiscus*, while he calculated that 4863 grains were produced by a single flower, or eighty-one times too many.† Mr. Darwin says, "In order to compensate the loss of pollen in so many ways, the anthers produce a far larger amount than is necessary for the fertilisation of the same flower; ... and it is still more plainly shown by the astonishingly small quantity produced by cleistogene flowers, which lose none of their pollen, in comparison with that produced by the open flowers borne by the same plants; and yet this small quantity suffices for the fertilisation of all their numerous seeds."

Mr. Darwin observed that when flowers were artificially self-fertilised for several successive generations, a degeneracy sometimes took place in the anthers and pollen; and he seems to attribute this to what he called the "evil effects" of self-fertilisation; but from the above-mentioned facts, which occur so abundantly in nature, I am inclined to regard it as an experimental verification and illustration of a universal principle in nature, namely the preservation of energy wherever possible, and that such cases as appeared under his

* *Cross and Self Fertilisation of Plants*, p. 25.
† *Ibid.*, pp. 376, 377.

experiments were instances of this principle at work, as the flowers became habituated to self-fertilisation, and were then fully fertile.

We have, then, in such cases an actual demonstration of the first step of the changes induced by self-fertilisation continually enforced; and thereby a witness to one cause of the origin of certain, and indeed, a very large number of species. It is the converse process to that of insect fertilisation, which *itself* I take to be the *vera causa* of the origin of intercrossing species.

It is, perhaps, worthy of note that, while both the number of stamens and the quantity of pollen are thus often much reduced in some flowers the capsules of which produce many seeds, yet in others which set but one, as *Fumaria*, or at least but few seeds, the number of stamens may remain unaltered. This seems to me to be an additional proof that such flowers are degradations from forms originally adapted to intercrossing when much more pollen was requisite. Hence the present forms are retentions of former ancestral conditions. The following cases will illustrate this:— *Scleranthus perennis* and species of *Medicago* have ten stamens and one seed; *Daphne Laureola* has eight stamens and one seed; *Chenopodium* has five stamens and one seed; similarly is it the case with the large orders *Compositæ* and *Gramineæ*.

The phenomenon called "contabescence" by Gärtner* would seem to have its rationale in this adaptation to self-fertilisation in some cases, and to diclinism in others, though there are other causes which may bring it about, when it is a purely pathological phenomenon.

Mr. Darwin observes, "The anthers are affected at a very early period in the flower-bud, and remain in the same state (with one recorded exception) during the life of the

* *An. and Pl. under Dom.*, ii., p. 165.

plant. The affection cannot be cured by any change of treatment, and is propagated by layers, cuttings, etc., and perhaps even by seed. In contabescent plants the female organs are seldom affected, or merely become precocious in their development. The cause of this affection is doubtful, and is different in different cases. . . . The contabescent plants of *Dianthus* and *Verbascum* found wild by Wiegmann grew on a dry and sterile bank." *

"Cases of an opposite nature likewise occur—namely, plants with the female organs struck with sterility, whilst the male organs remain perfect."

The constancy or prevalence of this condition of contabescence seems to be the first indication of diclinism, whatever the cause; and *Silene inflata* may be mentioned as frequently furnishing good examples of both kinds of contabescence.

DEGENERACY OF THE POLLEN.—As this is a feature of importance in the general degradation of flowers, a few words may be added in reference to it. It is of frequent occurrence in cultivated plants; thus Potatoes are notorious for failing to produce fruit; and some varieties are much less liable to do so than others. Mr. C. F. White, F.L.S., tells me he regards this plant as furnishing the most conspicuous example of a form of degradation of pollen; the pollen grains of a normal character are very generally not to be found at all, but round, square, and polygonal forms abound. On the other hand, he gathered many flowers, in a large field in the Isle of Thanet, with scarcely a grain imperfect in shape or reduced in size.

Mr. White has noticed, in his numerous researches among pollens, that degeneracy by dwarfing is mostly or very frequently induced by inclement weather. He mentions

* A like cause produces petalody of stamens, see p. 299.

DEGENERACY OF FLOWERS. 277

the case of "*Ononis*, growing and flowering abundantly on the 'Sand-totts' near Burnham, on the Bristol Channel, in which plant scarcely a grain of normal form was to be found; many were absolutely united into grotesque groups and utterly deformed.. At the commencement of the cold weather of autumn, although the corolla may appear uninjured, the pollen grains are often 'dirty,' unable, as it were, to throw off the residual tissue surrounding them, and are often irregularly reduced in size."

This sensitiveness of pollen to barren soil, inclement weather, etc., at once throws light on a probable origin of diclinism, such as of gyno-diœceous plants already mentioned; and simply confirms the idea that these differences in the sexual systems of plants must not be looked upon as so many beneficial arrangements, but simply inevitable results which must follow such circumstances as give rise to them, whether they may prove advantageous or not. The injurious effect of over-crossing, abundantly proved by florists, Mr. White recognizes in the character of the grains of Rhododendrons and Ericas, which exhibit a shrivelling up and occasionally a complete "dissolution" of *one* and the uppermost grain of the group of four. And this observer adds, that in more than one species of *Erica* and also of *Vaccinium* the injury, he thinks, has become chronic.

If the "vegetative" system be too energetic the "reproductive" is sure to suffer, and one of the primary causes of the injury is the arrested state of the pollen, as Van Tieghem has described and figured it in *Ranunculus Ficaria.** A like result occurs in many cultivated plants, as Mr. Darwin has pointed out when describing the "contabescence of anthers." ‡

* See above, p. 231, note.
† *An. and Pl. under Dom.*, vol. ii., p. 165.

278   THE STRUCTURE OF FLOWERS.

Degeneracy in the Gynœcium.—If the theory be true that a typical flower should contain two whorls of carpels, or, if spirally arranged, several cycles, then it is an obvious fact that these conditions are not the prevailing ones in nature. In a simple type, like *Ranunculus*, we find the pistil of many carpels, but with one ovule in each alone developed, except in monstrous conditions; if the ovules be numerous, then the carpels are reduced in number, as in the *Helleboreæ*. This is a primary result of Compensation. And when carpels have become whorled—a condition I take to be primarily due to adaptations to insect agency, causing an arrest of axial growth by the enhancement of the corolla, etc., (see p. 6)—then degeneracy begins to play an important part, in that, firstly, (theoretically, be it observed) one of the two whorls of carpels goes altogether, sometimes the calycine (e.g. *Fuchsia*), at others the petaline (e.g. *Campanula*).

Secondly, the number of carpels diminishes, as in the *Gamopetalæ*, where less than five prevail. The following table will show with tolerable accuracy the proportional number of carpels and ovules that prevail in the first three divisions of *Dicotyledons*.

	Thalam. Ord. p.c.	Calyc. Ord. p.c.	Gamop. Ord. p.c.
(1) Orders with many carpels or many ovules	12 or 19	6 or 7	0 or 0
(2) Orders with 5 carpels and many ovules	12 or 19	10 or 12	7 or 12
(3) Orders with 5 carpels and 5–10 ovules	12 or 19	14 or 17	3 or 5
(4) Orders with less than 5 carpels and less than 5 ovules	14 or 21	30 or 36	23 or 40
(5) Orders with less than 5 carpels and many ovules	17 or 25	22 or 27	25 or 43

Observations.—(1) The first-mentioned correlation has two

conditions, either many carpels having one or few ovules in each, or a few carpels with many seeds, as in the *Ranunculaceæ*. This primitive condition rapidly vanishes in passing to *Calyciflorœ* and *Gamopetalæ*.

(2) Having reduced the number of carpels to a definite quantity, five, *i.e.* one cycle of the prevailing ⅖ type, this number remains tolerably persistent, but does not show a large percentage.

(3) The combination of five carpels with a reduced number of ovules, *i.e.* one or two in each cell, or 5–10 ovules in all, is pretty uniform for the first two divisions, but almost disappears under *Gamopetalæ*, the orders *Sapotaceæ*, *Nolaneæ*, and one or two *Rubiaceæ*, (e.g. *Erithalis*) representing this condition.

(4) and (5). Here we see a steady increase in the percentages in passing from *Thalamiflorœ* to *Gamopetalæ*, in which the number of carpels is still further reduced; but the number of ovules runs in two directions, being either numerous or few.

Two questions arise at this point. If one result of insect agency is to bring about increased specialization in flowers (yet, in proportion as they become specialized, so, inversely, is the number and variety of insect visitors diminished), how is it that some (*e.g.* Foxglove and Orchids) produce an enormous number of seeds; while others (e.g. *Labiatæ, Compositæ*, etc.) produce few or only one in each flower? The second question is whether a plant is better off for having so many more seeds than another. Recognizing reproduction as the sole end of plant life, so that a plant should bear as many good seeds as possible, it is noticeable that the two largest orders, *Compositæ* and *Gramineæ* have never more than one seed to each flower. Again, comparing *Labiatæ* with *Scrophularineæ*, according to the *Genera*

Plantarum of Bentham and Hooker, while the former has 2600 species, the latter has only 1900. Lastly, comparing two orders with regular flowers and two carpels, *Boragineæ* has 1200 species, and *Solaneæ*, 1250; while the former order never has more than four seeds to a flower, in the latter they are numerous.

If it were possible, we should procure statistics as to the relative degrees of abundance in individuals of two kinds at any place where they thrive. Casual observations certainly have not led one to notice any such *proportional abundance* of the many-seeded plants as theoretically *ought* to exist if all their seeds germinated and grew to maturity; for I have calculated the number of apparently good seeds in a large plant of Foxglove, and found it was one and a half millions. If we take a typical case, that of Orchids, whose flowers are certainly of those most highly adapted to insect agency, it is now well known that the proportion of *seedlings* to *seed* is infinitesimally small. Mr. Fitzgerald speaks of a *Dendrobium speciosum*, which bore 40,000 flowers open at the same time; but though the plant was growing in the open air and was exposed to the visits of insects, *only one flower produced a seed pod.*[*] Mr. H. O. Forbes found the same thing to occur in the terrestrial orchids of Portugal, and the tropical ones of Borneo.[†] Exactly the same difficulties are met with in cultivating plants, and especially Orchids (with few exceptions), as Mr. Veitch has testified.

Now, when we examine the structure of the essential organs of Orchids microscopically, their degeneracy at once becomes apparent. First, with regard to the pollen. Instead of its being in well-formed distinct grains, each with its

---

[*] Referred to by Mr. Veitch, *Report on Orchid Conference*, Journ. Roy. Hort. Soc. Bot., vol. vii., p. 47.

[†] *Journ. Lin. Soc. Bot.*, vol. xxi., p. 538.

extine and intine, their development is arrested and, while still in contact, a common extine clothes the whole of each massula. Moreover, it is only after the pollen mass has been placed upon the stigma that the development is continued.* With regard to the pistil the first sign of degeneracy is seen in the parietal placentation which prevails, and more especially in the rudimentary character of the ovules, every part of which is degraded. Even after fertilisation the embryo cannot grow to maturity, but remains in the arrested pro-embryonic condition. Having no albumen or nucellus-tissue wherewith to nourish the embryo, the suspensor does its best by elongating and escaping from the micropyle, and then, fastening itself like a parasite upon the placentas, extracts nourishment therefrom—the result being that myriads of seeds never succeed (at least in cultivation) in developing even the pro-embryo; and one can only infer that such is the case in nature.†

In the cultivation of other flowers analogous phenomena are met with. The more highly cultivated a florists' flower may be, the less good seed is procurable; while the poorer ones—that is, from a florist's point of view—or "weedy" looking plants furnish plenty, and are highly prolific.

The rationale of these facts, whether taken from nature or from cultivation, I believe to be fundamentally the same, viz. the adaptation to insect agency and the result of repeated intercrossing, which enhances the development and form of the perianth especially, and generally of the stamens as well. At least the kinds of energy which are concerned in the manufacture of these whorls are more especially forced into activity by the stimulus received from without. On the other hand, the pistil suffers proportionately in all its parts

* Mr. B. T. Lowne, *Orchid Conference*, etc., *l.c.*, p. 48.
† M. Guignard has drawn similar conclusions. See above, p. 172.

through compensation and atrophy, the ovules being apparently particularly sensitive. To meet this difficulty nature seems, to speak metaphorically, to have tried two methods, either to make an immense number of seeds, so that at least a few might be perfect, or else to attempt no more than four or even one, so that at least they should be vigorous, and survive in the struggle for life during the critical periods of germination and seedling existence. To judge by results, this latter method turns out to be the best.

The interpretation, then, I would offer of inconspicuousness and all kinds of degradations is the exact opposite to that of conspicuousness and great differentiations; namely, that species with minute flowers, rarely or never visited by insects, and habitually self-fertilised, have primarily arisen through the neglect of insects, and have in consequence assumed their present floral structures. The external stimulus or irritations derived from the weights, pressures, and punctures of insects being no longer applied, the secretion of honey has failed, the corolla ceasing to be subject to hypertrophy has atrophied. A like procedure has obtained with the stamens, while a large proportion of pollen has become effete, the anthers being partly contabescent, as it is called. What remains, though often altered in character, is amply sufficient to set an abundance of seed.

With regard to the pistil, however, the reverse of this has in some respects taken place. The corolla and androecium no longer putting a check upon the rapid development of the gynoecium, the latter has a strong tendency to gain the ascendancy; so that the result is homogamy or protogyny, with an extraordinary fertility of all plants which have inconspicuous and regularly self-fertilising flowers.

If the seed be not always in great quantity in one and the same capsule, an ample progeny is secured by the

DEGENERACY OF FLOWERS.

extremely rapid maturation of the fruits in succession; as may be remarkably well seen in Chickweed.

The general result is that all these "weed-like" plants, with which wind-fertilised herbs must be associated as equally independent of insects, of all flowering plants are by far the most widely dispersed, and are, in fact, cosmopolitan;* and although they be small and annuals, are yet best capable of holding their own in the great struggle for life.

RUDIMENTARY ORGANS.—These are the ultimate result of atrophy and degeneracy in flowers. They are so well known as occurring in all parts of plants, vegetative and reproductive, that I need not describe them now. The reader will doubtless gather from all that has been said about hypertrophy and atrophy as causes of development and degeneration respectively, that they are just what one would expect to find. Indeed, every organ can be met with in every stage of degeneration till it has completely vanished; and even when all visible trace is wanting, the vascular cord belonging to it may in some cases still be detected. Last of all, this vanishes as well. These differences, for instance, can be witnessed in the presence or absence of the "trace" of the fifth stamen of the Labiatæ.

It is thought by some that a rudimentary organ may become a honey-secreting gland, as Robert Brown suggested for some Cruciferous plants. Glands mostly consist of epidermal and sub-epidermal tissues only, and if they occupy the place of an organ, the latter has the vessels arrested before they reach into the gland, which therefore is still of the same nature. In the male flower of *Lychnis dioica* the disk surrounds the rudimentary pistil, which in no way contributes

* In my essay referred to, I have given a long list of self-fertilising plants which have been discovered in widely distant localities over the northern and southern hemispheres.

to it. On the other hand, a gland may have its own proper vascular system, as in *Lamium album*, in which case a circular horizontal ring of vascular cords is formed from the pistillary cords; from this are given off a series of vertical cords, running up into the gland itself.

There can be no *à priori* objection to the supposition that an organ, when degenerating and becoming rudimentary, may acquire a new form and function; for such, indeed, is not infrequently the case. But what perhaps may be more usual, is that some other organ becomes more highly developed through compensation. Thus, for example, the leaflets of the Pea, in becoming tendrils, lose all trace of a blade, retaining only their mid-ribs. These, however, now elongate and acquire sensitiveness, for the use of climbing. On the other hand, in compensation for the loss of a certain amount of leaf surface, the stipules are very broad and foliaceous. Again, in the ray florets of *Centaurea* the essential organs have vanished altogether, but the corolla is greatly enlarged in comparison with those of the disk florets.*

* For a discussion upon "rudimentary organs," and their bearing upon the theory of Evolution, I would refer the reader to my work on *Evolution and Religion* (the "Actonian" Prize Essay for 1872), chap. xiii., p. 197.

CHAPTER XXVIII.

PROGRESSIVE METAMORPHOSES.

HOMOLOGY.—The theory of homology has long been maintained, and has met with such an overwhelming mass of evidence in its favour, that it is now regarded as a well-established morphological doctrine. The belief that every individual member of a flower, whether sepal, petal, stamen, or carpel, may be interchangeable with a leaf, and that they are therefore all phyllomes or foliar appendages to the axis, scarcely requires proof. Secondly, any one organ may theoretically be substituted for any other, so that although a sufficient number of interchanges has not yet been met with to make a complete series of permutations, yet they have gone far towards strengthening the probability that such might be possible.*

I propose giving a very abbreviated series to illustrate, first, progressive changes from leaves through bracts to

* The metamorphosis, with the exception of the substitution of petals for other organs, is rarely more than tentative; for it is, as it were, a mere attempt to effect a change, so that wherever a "monstrous" organ bears ovules they are almost always rudimentary and quite incapable of being fertilised. I have said "rarely," for M. Brongniart succeeded in obtaining fertile seeds from artificial impregnation of ovuliferous stamens in *Polemonium cœruleum* (*Bull. Soc. de Bot. Fr.*, t. viii., p. 453).

carpels; and, secondly, a retrogressive series from carpels to bracts, and thence to leaves; finally deducing some important conclusions.

PROGRESSIVE CHANGES IN BRACTS.—Bracts are in many cases very obviously modifications of leaves, being sometimes simply complete leaves reduced only in size, as in *Epilobium;* or a bract consists either of the blade alone, as in Buttercups, or else of the petiole only, but now expanded and blade-like in form, as may be well seen in Hellebores, where transitional states occur between the normal pedate leaf and true lanceolate bracts (Fig 61, *a, b, c*).

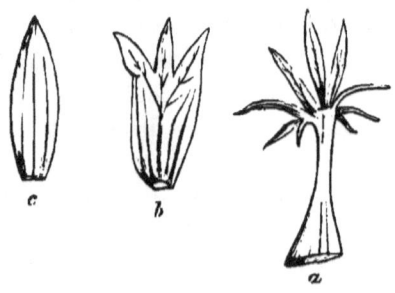

Fig. 61.—Transitional forms, *a, b*, from a leaf to a true bract, *c*, of *Helleborus viridis*.

When bracts are coloured otherwise than green, they then approach nearer to members of the reproductive or floral series rather than the vegetative, and in many cases are actually continuous in a spiral series with the sepals and petals, as in *Cactus, Calycanthus,* etc., and so assist in rendering the flower attractive. Several species of the genus *Salvia,* e.g. *S. splendens, S. Bruantii,* as well as of *Bromeliaceæ,* are remarkable for having brilliantly coloured bracts at the base of the flower. In some cases the bracts may be so arranged as to mimic a corolla, and indeed functionally replace it, as in species of *Cornus* (Fig. 62), *Darwinia* (Fig. 63), and the so-called Everlastings.

The presence of bright colours in bracts, as also in sepals, to be described, I take to be due to the same influence as of the normal attractiveness in corollas; viz., the visits of insects: the immediate cause being nourishment; the stimulus required to bring the extra flow to the bracts, etc., being presumably the irritation induced by insect visitors.

The next progressive state is for bracts to assume a more

Fig. 62.—Inflorescence of *Cornus florida*, with four white petaloid bracts.

Fig. 63.—Inflorescence of *Darwinia*, with coloured petaloid bracts.

or less staminoid character. This is rare, but it has been noticed in *Abies excelsa*.* A substitution of anthers for bracts has been seen in *Melianthus major*,† concerning which Sig. Licopoli remarks that the flowers of chiefly the terminal racemes were imperfect, the summit of the floriferous axis bearing a tuft of perfect and imperfect anthers the petals and the two carpels of the flower having been atrophied or arrested.

Fig. 64 represents an involucral bract of *Nigella*, bearing an anther on one side of it; while Fig. 65, *a*, is that of a glume of *Lolium perenne* with an anther. That bracts should ever assume a pistilloid character is, à priori, still more unlikely, as being further removed from the central organ of the flower. Dr. M. T. Masters has, however, described ‡ a

---

* *Teratology*, p. 192. † *Bull. Soc. de Bot. Fr.*, *Rev. bib.*, t. xiv., p. 253.
‡ *Journ. of Lin. Soc. Bot.*, vol. vii., p. 121.

malformed *Lolium perenne*, in which the flowering glumes had styles and stigmas (Fig. 65, *a*, *b*); the essential organs being absent, were replaced by a tuft of minute scale-like

Fig. 64.—Involucral bract of *Nigella*, with anther (after Masters).

Fig. 65.—Glumes of *Lolium*, with anther and stigmas (after Masters).

organs, some of which were prolonged into styliform processes, the sexual organs being otherwise suppressed.

In a proliferous case of *Delphinium elatum* described and figured by Cramer,* the parts of the flowers were all metamorphosed into open rudimentary carpels. The axis was elongated and terminated above, in one case, by a similar abortive flower; in another, by an umbel of such flowers, every part of which was more or less carpellary; while all the *bracts* on the prolonged axis, even those out of the axils of which the branches of the umbel sprang, were similarly made of open carpels.

PROGRESSIVE CHANGES IN THE CALYX.—The sepals are usually homologous with the petiole of a leaf. This is obviously the case with the Rose, where the rudiments of the

* *Bildungsabweichungen*, etc., heft. i., taf. 10. The figure is reproduced in *Teratology*, p. 126.

compound blades are retained (see Fig. 24, p. 93). In *Pedicularis* the blades are present as a minute fringe on the edge. In *Ranunculus, Potentilla*, etc., the broad base of the petiole is the only part present, for in abnormal conditions the blade may be borne above (Fig. 66). Similarly, in a gamosepalous calyx the teeth as a rule seem to be all that remain to represent the blades ; for in *Trifolium repens*, when virescent, true unifoliate blades are developed on elongated pedicels, all arising from the border of the calyx-tube (Fig. 67), in which the teeth become pinnately nerved blades.

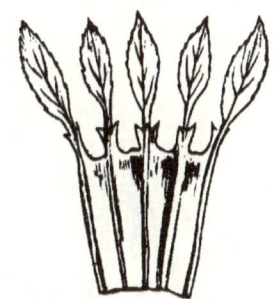

Fig. 66.—*Ranunculus* with foliaceous sepal.

Fig. 67.—Foliaceous calyx of *Trifolium repens*, with stipulate leaflets (after Baillon).

The venation may in some cases assist in furnishing a clue as to the real nature of a part. Thus in Hellebore, as already seen (Fig. 61), the bracts are homologous with petioles, their venation being palmate, and not pinnate as in the divisions of the blades of the leaves. It is the same in the sepals, which are presumably therefore homologous with petioles as well. The sepals of *Caltha* resemble them in their venation, but in this plant the leaf is of a more primitive type, not being lobed, and has also a palmate venation.

A similar difference between the venation of the sepals

and blades of the leaves is seen in *Dipterocarpus* and *Mussænda* (Fig. 68). Transitional states from a single to a double flower of *Saxifraga decipiens*, described and figured by M. C. Morren,* shows that the newly formed petals in the place of stamens, as also the normal petals of the flower, exactly correspond, both in shape and venation, with the cotyledons. Palmate venation thus simply represents a more primitive type; and, since flowers are constructed out of metamorphosed leaves—the vegetative being replaced by reproductive energies,—one naturally expects to find the calyx and corolla, which more nearly approach leaves in structure, to show arrested foliar conditions, as, *e.g.*, are seen in palmate nervation and absence of blade or petiole, as the case may be.

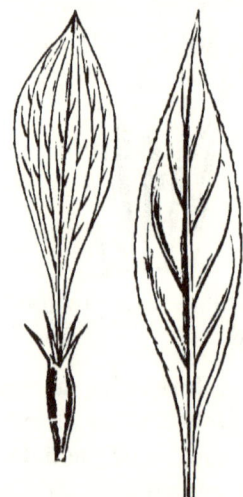

Fig. 68.—Flower and leaf of *Mussænda*.

In *Mussænda* (Fig. 68) the teeth of the sepals are usually subulate and acuminate; but in the one foliaceous and subpetaloid sepal it is drawn out into a long petiolar form, which then expands into a palmately nerved lamina. The fact that a "tooth" is in this case prolonged into a "petiole" seems to imply that the sepal arises at once from the receptacular tube, which, therefore, one would infer to be axial. A somewhat analogous procedure is in the monstrous *Trifolium*, where the unifoliate blades, supported on long pedicels *with stipular appendages as well*, all arise from the border of the so-called calyx-tube (Fig. 67). There the inference would be the same, only that the receptacular tube is free from the

* *Les Bull. de l'Acad. Roy. de Bruxelles*, t. xvii., p. i., p. 415.

pistil, and not adherent as in the case of *Mussænda*. In both instances it will presumably be purely axial in character.

Progressive changes in the calyx are not uncommon by its assuming a petaloid character. This is normal in some genera of *Ranunculaceæ*, in *Fuchsia*, *Rhodochiton*, as well as in some members of the *Incompletæ*, as in *Mirabilis*, *Polygonum*, *Daphne*, etc. Normally coloured sepals are most frequent in polysepalous genera. Abnormal colorisation, with or without any metamorphosis of the organ, is most frequent in gamosepalous flowers, as in the cultivated "hose-in-hose" varieties of *Primula*, *Mimulus* and *Azalea*. The calyx may be petaloid either wholly or in part only. In *Mussænda* (Fig. 68), one sepal only is normally sub-petaloid. *Calceolaria* has occasionally one or more sepals petaloid. Similarly *Linaria* (Fig. 69) and other instances might be mentioned. These conditions, brought about by cultivation, clearly show the important part that high nourishment plays as an external stimulus or factor in the production of colour.

Fig. 69.—*Linaria*, with one sepal petaloid.

Staminoid sepals appear to be very rare. It is recorded by M. Gris that they have occurred in *Philadelphus speciosus*:*

Pistiloid sepals are nearly equally as rare as staminoid. They have been observed by Mr. Laxton in double flowers of the Garden Pea (Fig. 70), in which there was a five or six-leaved calyx, some of the segments of which were of a carpellary nature, and bore imperfect ovules on their margins, the extremities being drawn out into sub-stigmatiferous styles.†

* *Bull. Soc. de Bot. Fr.*, t. v., p. 330.
† *Gard. Chron.* 1886, p. 897; and *Teratology*, p. 302.

292   THE STRUCTURE OF FLOWERS.

I have also found the sepals ovuliferous in a monstrous form of Violet, which was almost entirely virescent (Fig. 71).

PROGRESSIVE CHANGES OF THE COROLLA.—For petals to become staminoid is far from uncommon. It is a normal condition in *Atragene* (Fig. 44, p. 141), which illustrates the transition, and in Water-lilies, where a gradual development of the anther cells is accompanied by a gradual reduction of the petal to a filament. As abnormal instances may be mentioned, a case of Foxglove which I have elsewhere * described as having the corolla split up into strap-shaped antheriferous processes (Fig. 72), and a Columbine in which the spurs

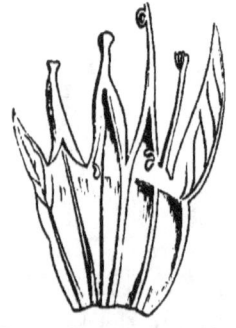  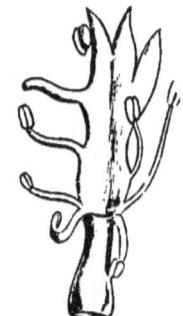

Fig. 70.—Calyx of Garden Pea, with carpellary lobes (after Masters).   Fig. 71.—Ovuliferous sepal of Violet.   Fig. 72.—Corolla of Foxglove, with staminate tube.

became curiously coiled and bore pollen within the tissue of the coils (Fig. 73).

Pistiloid petals are of rare occurrence. As an example is *Begonia* (Fig. 74, *a*), in which the apex of the petal was green and stigmatiform, the basal part being broad, coloured, and ovuliferous. Fig. 74, *b*, shows a petal, ovuliferous below, stigmatiferous at the summit, and antheriferous midway; *c* is a rudimentary ovule.

PROGRESSIVE CHANGES IN THE STAMENS.—The only change

* *Journ. Linn. Soc. Bot.*, vol. xv., p. 86, tab. 3.

that stamens can undergo in this direction is to be more or less converted into pistillary structures. This is by no means uncommon. Either the filament alone, or the anther alone, or both together may be affected. The reader is

Fig. 73.—*Aquilegia*, with polleniferous spurs (after W. G. Smith).

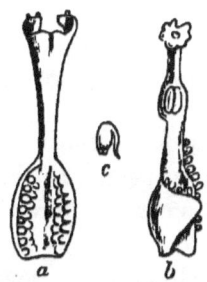

Fig. 74.—Ovuliferous petals, etc., of *Begonia* (after Masters).

referred to Dr. Masters's *Teratology* for a description, with figures of several kinds.* It is more usual for the filament to become enlarged into the ovarian part bearing rudimentary ovules; but when the anther is involved, it may be partially or wholly transformed. In these cases the connective is usually prolonged into a stigmatiferous process.† As an example often described is that of the Houseleek, in which the margins of the anther cells become ovuliferous in various degrees; as in Fig. 75, where ovules are borne by the posterior sides only, instead of pollen. In other cases the filament bears rudimentary ovules as well. Dr. Masters points out that "where there is a combination of the

* Page 303.
† In *Aristolochia* this change seems to be permanent and functional. See above, p. 83.

attributes of the stamen and of the pistil in the same organ, the pollen is formed in the upper or inner surface of the leaf organ, while the ovules arise from the opposite surface from the free edge." *Begonia* is a genus which is peculiarly liable to produce malformations in the stamens (Fig. 76).* *Rosa*

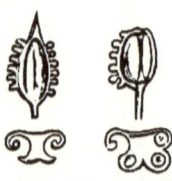
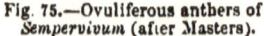
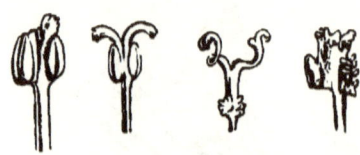

Fig. 75.—Ovuliferous anthers of *Sempervivum* (after Masters).

Fig. 76.—Stigmatiferous and ovuliferous stamens of *Begonia*.

arvensis † affords a case in which the ovules were borne by the anthers, and then they themselves produced pollen. In these cases, where the anthers are ovuliferous, the connective is often more or less stigmatiferous, as in *Begonia* (Fig. 76), which shows various degrees of metamorphosis in this way; but the anthers may sometimes be stigmatiferous, as in Poppies, ‡ or styliform as well, as in Bamboos. §

The complete substitution of carpels for stamens occurs in many plants, as in *Malus apetala*,|| Tulips, etc., and is extremely common in Wallflowers,¶ while it is by no means an uncommon occurrence to find male plants of normally diœcious or monœcious character bearing female organs, though perhaps in these cases it is often an *addition*, rather than a *substitution* of one organ for another.

* See *Journ. of Lin. Soc.*, xi. 472; *Bot. Zeit.* (1870), vol. xxviii., p. 150, tab. ii.

† *Journ. of Bot.*, 1867, p. 318, tab. 72.   ‡ *Teratology*, p. 304.

§ Col. Munro, *Trans. Lin. Soc.*, vol. xxvi., p. 7.

|| Poiteau et Turpin, *Arbr. Fruit.*, t. xxxvii., referred to by Moquin-Tandon, *Tératologie*, p. 220.

¶ Called "Rogues" by the market-gardeners, as the corolla is wanting or green. See *Ann. des Sci. Nat.*, 5 sér., xiii., p. 315, pl. 1.

CHAPTER XXIX.

RETROGRESSIVE METAMORPHOSES.

THE PISTIL.—Commencing with the pistil, there may be changes in the ovary, ovules, style, and stigmas, separately or collectively. Instead of one or more ovules, a pistil may be formed within an ovary, as sometimes occurs in Wallflowers, Grapes, Oranges, etc.* A singular instance is described by Dr. Masters † of a Carnation, "the placenta of which bore not only ovules but also carpels, the latter originating in a perverted development of the former; so that many intermediate stages could be traced between the ordinary ovule and the ovary. Some of these carpels, thus derived from ovules, themselves bore secondary ovules on a marginal placenta" (Fig. 77, a, carpel and section), the secundine, however, being the only part developed (b).

Fig. 77.—Carpels and ovules on placenta of Carnation (after Masters).

Stamens within an apparent ovary have occurred in

* *Teratology*, p. 182.
† *L.c.*, p. 267. Perhaps the supposed "ovule within an ovule" may have been the nucellus only, more or less free from the secundine.

Bæckia diosmæfolia; * but as they grew on the interior of the wall and not on an axile placenta, as is the normal condition in the *Myrtaceæ*, I expect that it was due to the staminal vascular cords branching off and coming out of the tissue *within* instead of at the summit of the hollow receptacular tube, the carpels being more or less arrested. A not uncommon instance is to find the pistils of Willows with open ovaries and bearing one or more anthers on the margins (Fig. 78, *a*). I have met with a similar occurrence in *Ranunculus auricomus* (Fig. 78, *b*). Pistils of other flowers

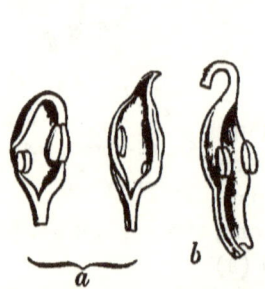
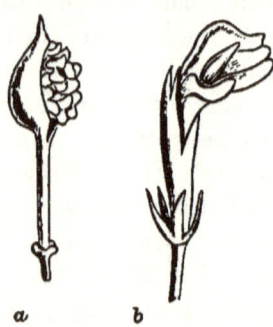

Fig. 78.—Stameniferous carpels of Willow (*a*) and *Ranunculus auricomus* (*b*).

Fig. 79.—*a*, Petaliferous placentas of *Cardamine pratensis*; *b*, of *Rhododendron*.

have been known to bear anthers in a similar way, as *Chamærops humilis, Prunus,*† etc.

Pollen within ovules has been met with occasionally, as in *Passiflora* and *Rosa arvensis*.‡

In some members of the *Cruciferæ*, as *Cardamine pratensis* (Fig. 79, *a*), round pods are formed instead of the usually

* *Teratology*, p. 184. Possibly the ovary was entirely absent, and the stamens would then be growing on the interior of a closed receptacular tube, just as carpels grow upon the inside of the hip of a rose.

† See Weber, *Verhandlung des Nat. Hist. Vereines der Preuss Rhein. und Westph.*, 1860, p. 381.

‡ *Teratology*, p. 185.

long siliquas. These are full of petals, and if carefully examined appear to be whorled, with traces of stamens and pistil within them; so that they represent flower-buds, but of which petals form the greater part; similarly, Rhododendrons and other flowers are known to bear imperfect flower-buds within the ovary (Fig. 79, *b*).

Anthers occupying the place of stigmas appear to have occurred in Campanula,* Snowdrop, and double Tulips.

The substitution of stamens for the entire pistil is of a less usual occurrence than the staminody of its parts: for cases, the reader may consult Masters's *Teratology*.† In a species of *Orchis*, probably *O. Morio*, the ovaries were wanting altogether, a long pedicel taking their place, and within the reduced and regular perianth were *two* anthers on opposite sides (Fig. 23, *a*, p. 92), an apparent compensation in lieu of the pistil.

The next and most frequent case of metamorphosis is that of conversion of carpels, and usually the stamens as well, into petals, or the so-called "doubling" of flowers. This is usually accompanied by a change from whorls to spirals with a multiplication of the parts. Thus, in a double Wallflower, I have counted more than fifty petals spirally arranged. With regard to the petalody of the pistil, as Dr. Masters observes, "this is much less common than the corresponding change in the stamens. It generally affects the style and stigma only, as happens normally in *Petalostylis*, *Iris*, etc." ‡ Fig. 80 illustrates a metamorphosed carpel of *Polyanthus*, with a broad coloured appendage to the style.

Fig. 80.—Metamorphosed sub-petaloid carpel of *Polyanthus*.

In some double flowers the carpels only are petaloid.

* *Teratology*, p. 300.   † *Ibid.*, p. 299.   ‡ *Ibid.*, p. 296.

This has been observed in *Anemone nemorosa*, cultivated varieties of *Ranunculus*, *Violet*, and *Gentiana Amarella*.

RETROGRESSIVE METAMORPHOSES OF STAMENS.—For the stamens to become petaloid, it is extremely common, as in double flowers, and such a change may represent what is normally the case in Water-lilies, *Canna*, and *Atragene* (Fig. 44, p. 141). Changes may apply to the anther lobes, connective, or filament, or to all together. Fuchsias often bear filaments with petaloid expansions of the apex, at the base of which are one or two anthers showing varying degrees of degeneration. This is a very similar condition to one in *Petunia*, described by Dr. Masters, in which the connective had developed into a green roundish blade bearing two anther cells at the base (Fig. 81).* In such cases, it seems to be the connective which has expanded outwards and become the blade of the petal or leaf. Similarly, in the

Fig. 81.—Foliaceous connective of *Petunia* (after Masters).   Fig. 82.—Petalody, or "hose-in-hose" form, of connectives in a double Columbine.

double Columbine petalody of the connective sometimes takes place (Fig. 82).† *Commelina alba* has also furnished a case of an anther lobe becoming petaloid.

CAUSES OF "DOUBLING."—There can be no doubt that petalody results from a weakened reproductive energy, especially that of the androecium, which can become constitutional and may be hereditary and transmissible by crossing.

* *Teratology*, p. 254.   † *Ibid.*, p. 293

Cases seem clearly to show that a barren and dry soil, as well as a very dry atmosphere, are prominent causes for its appearance. Thus Mr. Darwin described a double *Gentiana Amarella*,* growing "on a very hard, dry, bare, chalky bank." T. S. speaks † of a double *Potentilla* as "growing along a high wall, on a dry raised bank close to a beaten path, adjoining a gravelly field." Again, a writer in *Gartenzeitung* ‡ alludes to the raising of double Stocks, and says that they should only have "just enough water for their preservation," and that "the starved state of the plants" causes doubling. He alludes to Camellias, also, as becoming double when grown in a dry soil. *Kerria Japonica* becomes double in Europe, in consequence of its missing the wet season of Japan. It is well known that double flowers are more easily raised on the continent than in England, probably from a like cause, as our atmosphere is considerably more charged with moisture than a continental one. In raising double Stocks, it is customary to procure seed from the flowers on axillary shoots which have a weaker reproductive energy than those growing on the primary or central axis, the seeds being smaller and often misshapen. The above causes are, therefore, suggestive; in that if a somewhat elevated, dry, and poor soil, one devoid of phosphates, etc., be provided, the probability is that petalody will ensue. Having once shown a trace of the malady, florists know how to proceed in order to propagate and transmit the affection.

There remains one other floral metamorphosis, and that is of petals into sepals. This condition approximates to virescence of the corolla, so that in many cases such a change could scarcely be called sepalody. But M. Godron has shown that when *Ranunculus auricomus* appears to be

* *Gard. Chron.*, 1843, p. 628   † Ibid., 1866, p. 973.
‡ Ibid., 1886, p. 197.

apetalous or to have a corolla consisting of a few petals only, it is due to the fact that the petals which are wanting are really present, but have become calycine.

ORIGIN OF HOMOLOGY.—Though we cannot penetrate into the arcana of life, nor trace the workings of its forces which bring about the development of any organ whatever, I think we can at least reach the physiological starting-point, so to say, of all these changes which I have briefly described. I have already mentioned that we may consider a vascular cord as the fundamental "floral unit," and as all cords are identical in character as long as they are within a pedicel, and, as far as one can observe, identical also in character even when they have penetrated the different organs, we at once see that there is a common source for each and all. Secondly, when we trace these cords from the receptacle or axis into the floral members, we soon discover that any cord can supply two, three, or more totally different organs with their respective branches, as in the case of *Campanula medium* described above (p. 43). Indeed, starting, say, with five cords in a pedicel, they can supply any number of organs *ad libitum*, however diverse in character and however numerous they may be. Hence, although normally each whorl is stamped with its own individuality, it is easy to imagine, in accordance with the principles of evolution, that others may partake of it; and so the characteristic features peculiar to one whorl can transcend its limits, and influence others as well.

Beyond some such interpretation as this, I do not think it is possible to go.

In saying that a fibro-vascular cord can "give rise" to a sepal, or petal, or other organ, I need hardly remind the reader that I am only speaking metaphorically, in describing what one observes in studying the anatomy of flowers.

CHAPTER XXX.

PHYLLODY * OF THE FLORAL WHORLS.

VIRESCENCE AND FOLIACEOUS CONDITIONS—SEPALS, PETALS, AND STAMENS.—The last changes to be described, which are common to all the members of a flower, are virescence, when they retain their normal forms, but are simply green; and foliaceous conditions, when they assume more or less a truly leaf-like form.

Dr. Masters has given descriptions † of several of each kind of floral members as well as of foliaceous bracts, to which I must refer the reader for details. There are certain particulars, however, to which I would especially draw attention as throwing light upon the ordinary structure of floral whorls, and especially that of ovules.

Taking the Alpine Strawberry as an illustrative case, the petals, stamens, and carpels are often more or less foliaceous; but the petals retain a palmate venation, though the three leaflets of the ternate leaf are pinnately nerved (Fig. 83, a, b). In the case of stamens the connective may be foliaceous, as in *Petunia* (Fig. 81); ‡ also in the Alpine Strawberry (Fig. 83, a) and in the "Green Rose" the anthers are often persistent on either edge of a leaf-like intermediate part

* The abnormal assumption of a leaf-like character.
† *Teratology*, p. 241, seqq.
‡ *Ibid.*, p. 254.

(Fig. 83, c). A curious foliaceous modification is described by Müller and figured by Masters,* in which the metamor-

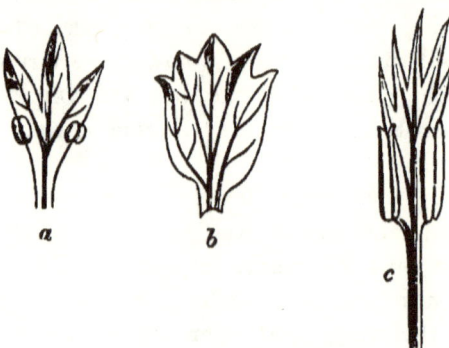

Fig. 83.—a, Foliaceous stamen, and b, petal of the Alpine Strawberry (after Le Maout and Decaisne); c, stamen of "Green Rose."

phosed stamen had the appearance of two leaves united by their mid-ribs. It occurred in *Jatropha Pohliana* (Fig. 84). This will be alluded to again, as peculiarly significant.

PHYLLODY OF THE CARPELS AND OVULES.—This is of much more frequent occurrence than of the stamens. The first condition of change is to leave the ovary open and to expose the ovules; the style may still be stigmatiferous. The ovules then undergo phylloidal changes of different degrees, and much discussion has arisen as to whether the coats of ovules should be regarded as homologous with leaves, the nucellus being axial, or not, etc.†

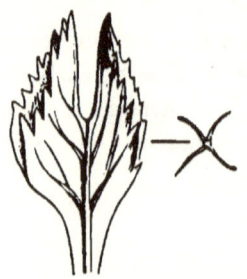

Fig. 84.—Foliaceous stamen of *Jatropha Pohliana* (after Masters).

Since, however, anatomical observations clearly prove that both the primine and secundine issue out of tangential

* *Teratology*, p. 255. † *L.c.*, p. 262.

divisions of the epidermal layer of the nucellus, one can hardly consider these of themselves as homologous with a true phyllome. But when we find that each of the two sides of an anther cell can develop into a foliaceous structure, as in the case of the *Jatropha* alluded to above (Fig. 84), we seem to have discovered a power of converting what is originally and simply an *epidermal layer* into a truly *foliaceous structure*. Moreover, this process is not infrequent in certain monstrous states of ovules, so that it would appear that any question of homology is, strictly speaking, out of court in these cases. When the whole of an appendicular organ becomes foliaceous, then, of course, a true case of homology may be recognized.

ORIGIN, DEVELOPMENT, AND HOMOLOGIES OF OVULES.—Teratology, here, I think, assists us greatly. With regard to the structure of an ovule, it first appears as a papilla upon the placenta, the cellular tissue of which, with its epidermal layer, constitutes the first stage. Such may, perhaps, be considered as the rudimentary condition of the funicle alone, as the true ovule is formed at the summit of it. One or more of the apical sub-epidermal cells gradually develop into the nucellus, while the secundine is first formed by tangential division of the epidermis commencing at a certain place below the apex; the primine, if present, subsequently following suit in the same manner.* It is a noticeable fact that while an ovule thus complete is elsewhere general in flowering plants, the Gymnosperms and most orders of the *Gamopetalæ* form remarkable exceptions, as having only one coat to the ovule. In the former of these two groups it is doubtless due to a primitive condition being accompanied by other features showing affinities with cryptogams. In

* See paper by Warming, *De l'Ovule*, Ann. des Sci. Nat., v. (1877), p. 177.

the latter, it is due to reversion by arrest, and is likewise accompanied by a simpler origin of the nucellus and embryo-sac, as Warming has shown. The suggestion I would offer to account for this anomaly is that such arrest is due to compensation. The *Gamopetalæ*, as a whole, are the most advanced of all flowers through adaptations to insect agency; and as this invariably brings about an exalted condition of the corolla and stamens, the consequence is that the pistil has to suffer; the first visible result being protandry, accompanied by a temporal arrest in the development of the pistil.

If this tendency to arrest be carried to the ovule, it may be affected too, and the result is that one, the last-formed coat, may be arrested altogether. Orchids, as shown above, illustrate this principle remarkably well, as their ovules, though possessing two coats, are as degenerate as in many parasitic plants (see above, pp. 172 and 281).

Tracing the origin of an ovule, then, from its birth, it first appears as a papilla on the placenta of the carpel. A branch from the marginal fibro-vascular cord of the carpel enters it from below, and reaches at least as far as the chalaza, or base of the nucellus. It may go no further, as in rudimentary ovules of *Orchis;* or be arrested in the form of cambium in the degraded state seen in the parasitic *Thesium*. On the other hand, as the ovule becomes a seed and the coats go to form the seed-skin, fibro-vascular branchings may occur all through the latter, being developed from the original cord. Such may be well seen in Mustard, Acorns, Beans, and the Coco-nut.

Although the coats of the ovule were originally formed by tangential division of the cells of the epidermis of the nucellus, when united to form a seed-skin, this has become thickened by a cellular growth between them, through which the cords then ramify.

PHYLLODY OF THE FLORAL WHORLS. 305

Pistils which have reverted to a more or less foliaceous character bear ovules which often become foliaceous as well; and then a not uncommon procedure is the development of a cup-like structure, probably composed of the two ovular coats, on an elongated stalk, with a rudimentary nucellus within, but more or less perfectly free from it; or it may not exist at all.

The late Professor Henslow described a monstrous condition of Mignonette with figures of ovules in this condition.*
They were sometimes replaced by minute leaves (Fig. 85, c); or else in the place of each was a cup-like structure, elevated on a long stalk, with an egg-like nucellus within, but quite free from it. He likened it to the theca of a moss with its central columella.

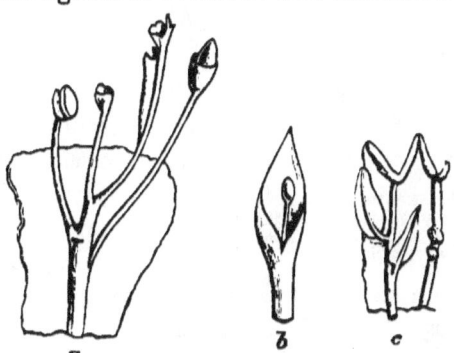

Fig. 85.—Foliaceous and metamorphosed ovules of Mignonette (after Prof. J. S. Henslow).

Comparing these two modifications, represented by Fig. 85, a and b with c,—or, again, those of Fig. 86, a and b,—the interpretation seems to be that the fibro-vascular cord passing up the funicle of the ovule becomes a petiole, and its prolongation constitutes the mid-rib. The secundine and primine with intermediate tissue become the blade, as seen in the foliaceous states of ovules, and constitute the "cup" when they assume that form.

A similar process, I think, quite explains the origin of the foliaceous processes of the stamen of *Jatropha*, represented by Fig. 84. The entire stamen is, of course, really

* *Trans. Camb. Phil. Soc.*, vol. v.

homologous with one leaf alone; but the membranes belonging to each anther, being of, at least, two layers of cells, have become foliaceous, just as the epidermis of the nucellus has done in the cases herein described; so that, in the *Jatropha*, two leafy expansions were developed out of one.

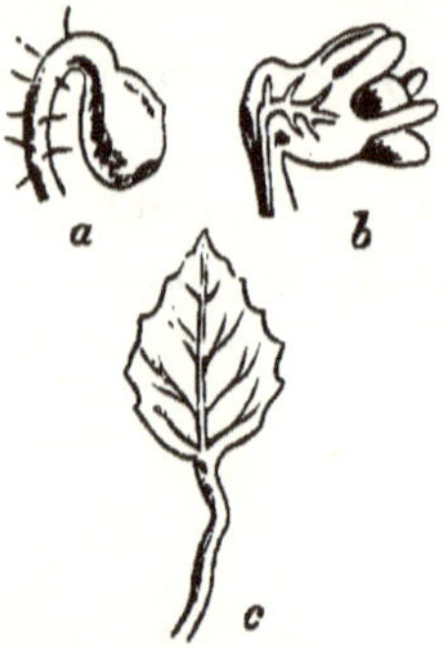

Other instances are known of ovules being represented by leaves, as *Primula Sinensis*, *Symphytum officinale*,* and *Sisymbrium Alliaria* (Fig. 86).

Theoretically, it might be objected that a leaf (carpel) should give rise to a leaf (ovule or, at least, ovular coat); but foliaceous excrescences from a leaf-surface are not at all uncommon, as, for example, frequently occur in Cabbages,† where, in consequence of high nourishment inducing hypertrophy, any "rib" or "vein" may throw off a branch which can form a leafy expansion, which not at all infrequently becomes funnel-shaped, like the abortive ovules of the Mignonette. Similar funnel-shaped or tubular productions are found on corollas of semi-double flowers, as in Primulas, Cyclamens, *Antirrhinum*, etc., sometimes externally. Fig. 87 represents a like outgrowth from the labellum of *Cattleya Mossieæ*; and I have seen the posterior sepal of *Vanda cœrulea* replaced by a pedicel with a cup at the apex exactly like the terminal process in Fig. 85, *a*. In all these cases I would regard such productions as due to hypertrophy.

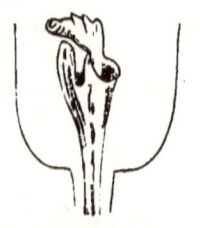

Fig. 86.—Metamorphosed ovules with foliaceous coats of *Sisymbrium Alliaria* (from *Linnæa*, xvii., t. 20).

Fig. 87.—Tubular excrescence on the labellum of *Cattleya Mossieæ*.

* *Teratology*, p. 263. † *Ibid.*, p. 312, fig. 166.

As another curious instance a remarkable form of the Sun-dew, *Drosera rotundifolia*, may be alluded to here, as throwing additional light upon the origin of ovules. It has been described and figured by Naudin,* and also by Planchon.† In this monstrosity the ovular coats were represented by "tentacles." These, as is well known, are not epidermal trichomes, but structures issuing from branches arising from the fibro-vascular cords of the leaf, and are therefore strictly homologous with the "funnels" on cabbage leaves.‡

The conclusion, therefore, which seems deducible from the foregoing observations is that an ovule is simply an appendage (not a bud) to the fibro-vascular cord of the margin of the carpel, and under monstrous conditions can grow into foliaceous excrescences to the carpellary leaf. It is not, therefore, axial in its character. Since all that is required to start from is a fibro-vascular cord, this may be furnished by any cord, even the mid-rib; and such is the case in some monstrous states of *Primula*, in which rudiments of ovules are found on the mid-ribs as well as on the margins of separate carpels.

As the "funnels" on the mid-ribs and lateral veins of cabbage leaves are due to an abnormal condition of hypertrophy, so ovules I consider as arising in a similar way, and take them to be due to the same influence, though of course it is normal in their case. The very presence of the large cords running up the margins of carpellary leaves, direct from the axis below,—being often, indeed, larger than the dorsal cord,—which then ramify, not only into each ovule, but often backwards within the carpellary walls till they reach and

* *Ann. des Sci. Nat.*, 2ᵒ sér., vol. xiv., p. 14.
† *Ibid.*, 3ᵉ sér., vol. ix., p. 86, tab. 5, 6.
‡ The "pitchers" of Nepenthes, perhaps, originate in much the same way, from the original water-gland at the apex of the leaf.

anastomose with the dorsal cord; these, together with the greatly thickened cellular margins now constituting the placentas which supply the conducting tissue for the pollen-tubes, all show a form of hypertrophy in the edges of the carpellary leaves, a condition of things widely different from the usually thin and more or less impoverished margins of true leaves.

If we may recognize a fibro-vascular cord as the "fundamental unit," and as a basis for the construction of any organ, and moreover as also containing within it potentially the power of evolving any number of similar organs by repeatedly branching; then, when hypertrophy affects such a "unit," it may branch once, twice, or any number of times, when each branch passing off to the surface can lay the foundation of a repetition of the organ from which it takes its rise.*

Attention has already been called to this origin of the numerous stamens of the *Malvaceæ*, and how certain forms of double flowers originate from the multiplication of the petaline cords, each branch of which issues in a distinct petal, as in Snowdrops.

Fig. 88.—Multifold carpels with ovuliferous margins from a malformed Primrose.

Similarly for carpels and ovules, the process of multiplication can be witnessed both normally and abnormally. On the one hand, that of carpels into five groups occurs in the Hollyhock through the chorisis of the original carpellary cord; on the other, Fig. 88 represents a multifold carpel of a Primrose, due, there is no doubt, to a like chorisis of the cords belonging to one individual carpel.

Similarly for ovules, while two only are normally charac-

* I must again remind the reader that I am here speaking metaphorically; as we do not know wherein this potentiality really lies, but can only describe what is actually visible.

teristic of the Plum, and Orchis has an innumerable quantity arising by repeated chorisis of the original placentary cords, so in monstrous Primroses, etc., as represented in Fig. 88, many additional ovular processes may be formed, not only on the margins, but even, as stated, on the mid-ribs.

One other point may be here noticed, *à propos* to the following curious discovery by M. Baillon. I have regarded a vascular cord as a "unit," as being capable of giving rise to any appendage whatever; and as long as it is in an axis as a "trace," the cords of all organs are absolutely indistinguishable. Further, there is no difference between a cord which will enter an appendage and one which will form a pedicel from a peduncle. In the latter case, several cords are usually required for the pedicel; while one, the most external of the "horseshoe" group given off at one side of the peduncle (*i.e.* as seen in a transverse section), enters the bract. In *Erodium cicutarium*, which has three flowers to the umbel with very slender pedicels, one single cord is all that the peduncle contributes to supply each of the pedicels, and one very small cord for a bract. The cord for the pedicel increases by radial chorisis, and so passes from the form of a wedge to that of a fan, when the outermost parts increase till they meet, and so a circle is established.

This shows that an "axis" and an "appendage" are fundamentally due to the same kind of "unit."

The reader will now see that in the following case * the funicular cord, which is normally that of a foliar, *i.e.* appendicular organ, supplied an axial cord; just as many leaves can give rise to buds which are often utilized for propagative purposes.

* *Sur le Développement et la Germination des Grains Bulbiformes des Amaryllidées*, par. M. Baillon, Bull. Soc. de Fr., t. xxi., 30 (pub. en *Revue des Cours Sci.* Lyon, Août 30, 1873).

"Les bulbilles des Amaryllidées ne sont pas toujours des graines véritables modifiées seulement quant à l'épaisseur et à la consistance de leurs diverses couches naturelles, notamment des plus extérieures. Témoin le *Calostemma Cunninghamii*. Ici, par une singulière transformation de l'ovule en bulbe, la chalaze, en s'épaississant, joue le rôle d'un véritable plateau, sur lequel se produisent une, puis plusieurs racines adventives. Les enveloppes ovulaires tienne alors lieu d'écailles bulbaires tandis qu'il s'élève dans le sac embryonnaire un véritable bourgeon parti de la chalaze comme support et s'échappant par son sommet de la cavité ovulaire pour se comporte ensuite comme une plante complète."

An analagous case of bulbs arising from a foliar organ occurred with *Scilla Sibirica*. Some plants dug up in October, 1887,* were found to have taken the form of the so-called "droppers" not uncommon in tulips. Their peculiarity resides in the fact that the tubular leaf-sheath bends and grows downwards, thereby carrying the axillary bulbil to a greater depth in the soil than usual. In February, 1888, on re-examining them, Dr. Masters discovered that from one to four bulbils had been developed at different heights within the tissue of the tubular sheath, being in connection with the cords of the latter by means of a transverse nexus of tracheids.

I refer to these cases as being curious instances of axial structures proceeding from foliar—*i.e.*, by the change of character of the fibro-vascular cords from being at first foliar and then axial. They support the theory of homology of leaf and axis, which is, of course, otherwise quite efficiently substantiated by such plants as *Xylophylla*, *Ruscus* and the *Cactaceæ*.

* *Gard. Chron.* for Oct. 15, 1887, p. 475, fig. 98, "droppers;" also, for March 3, 1888, p. 276, fig. 45, ditto, with bulbils.

CHAPTER XXXI.

THE VARIETIES OF FERTILISATION.

THERE are at least seven kinds of union :—(1) self-fertilisation, or the fertilisation of a pistil by the pollen from the same flower (Autogamy); (2) crossing different flowers on the same plant; (3) crossing flowers on different plants of the same stock; (4) crossing flowers of different plants, but of different stocks; all the preceding being of exactly the same form or variety of species; (5) crossing varieties of the same species; (6) crossing different species of the same genus; (7) crossing different genera of the same order.

When a knowledge of the floral sexes was first acquired, the idea maintained was that hermaphrodite flowers were specially adapted for self-fertilisation; but it was, I believe, Dean Herbert who first observed the importance of crossing, in his work on the *Amaryllidaceæ* (1836). He says, "I am inclined to think that I have derived advantage from impregnating the flowers from which I wished to obtain seeds with pollen from another individual of the same variety, or, at least, from another flower rather than its own, and especially from an individual grown in a different soil or aspect."

Mr. Darwin's work, *On the Cross and Self Fertilisation of Plants* (1876), placed on a scientific basis, by means of experimental verifications, the exact values of such crossings. His conclusions, however, require considerable modifications.

They are true for at least the first few years; as in but four or five cases only did he exceed the third generation; and when he prolonged them to seven generations, as in *Mimulus luteus*, and ten in *Ipomœa purpurea*, his results began to assume a very different complexion.

The inference deducible from his experiments is that careful and artificial crossing generally introduces a remarkable stimulus *for a time;* but the effects are not permanent. On the other hand, a perseverance in self-fertilisation produces results which are much more stable; so that, finally, self-fertilised plants (*i.e* the successive off-spring of this process) outstrip their competitors. Florists also find that by continued crossing the flowers of a species they soon reach the end of their tether, and no further progress is obtainable.

Secondly, Mr. Darwin failed to realize the fact that self-fertilisation predominates in nature with the vast majority of hermaphrodite plants, whether they be adapted to insects or are inconspicuous and adapted for autogamy.

Thirdly, he misinterpreted the meaning of degeneracy, which often accompanies self-fertilisation; thinking that it involved constitutional injuriousness, of which there is no trace whatever in nature.

Lastly,* he, and other writers who have followed him, wrongly inferred that adaptations to insect agency implied a converse "purpose," viz. to avoid self-fertilisation, instead of regarding them as *the inevitable results* of the stimulus of intercrossing and of the visits of insects. The danger of this à *priori*, deductive, or teleological reasoning, without any attempt at verification, lies in the fact that it is untrust-

---

* I must refer the reader to my paper on *The Self-fertilisation of Plants*, in which I have dealt with these points. It was written in 1877, but I have met with nothing since to invalidate the above conclusions, but, on the contrary, very much in support of them.

worthy. It may or may not be true; but it is of no value unless thoroughly tested by experiment and verified. Thus, for example, Mr. Darwin, in speaking of the movements of the stigmatic lobes of *Mimulus*, says "Mr. Kitchener has ingeniously explained the use of these movements, namely, to prevent the self-fertilisation of the flower." He, however, experimented with this plant, and then discovered that "if insects are excluded the flowers fertilise themselves perfectly, and produce plenty of seed."*

Again, it has been argued that we are justified in assuming that the remarkable adaptations to insects, which are so obvious in many flowers, *must be of some use* to the plant, even though we may not be able to discover it. This statement, however, is just as much an *à priori* and deductive assumption as the preceding, and is quite valueless until verified; and it is only by means of such experiments as Mr. Darwin laboriously carried out, that the real value of intercrossing and self-fertilisation or other kind of union can be ascertained. Thus, *e.g.*, the Garden Pea is undoubtedly adapted to insects, like other irregular flowers; but experiments proved that "a cross between two individuals of the same variety does not do the least good to the offspring, either in height or fertility."†

* *Cross and Self Fertilisation*, p. 64. As another instance of an *à priori* deduction, Sachs says of *Epipactis latifolia*, "The flower left to itself does not get fertilised, for the pollen-masses do not spontaneously fall out of the anther; and even if they did, would not come on to the stigmatic surface" (*Veg. Phys.*, p. 796). Mr. A. D. Webster, however, has observed that *E. latifolia* is very imperfectly fertilised, for, although visited by insects, cross-fertilisation seldom takes place; that "self-fertilisation by the pollen falling spontaneously on the stigma is not uncommon, as the pollen-masses ... become friable, and before the plant withers, either spontaneously or by the action of the wind fall on the stigma" (*Bot. Gaz.*, xii., p. 104).

† *L.c.*, p. 264.

Moreover, however greatly we may feel impressed with the truly wonderful adaptations of flowers, a careful and critical study of them reveals many features which seem to counterbalance, to some degree at least, the "good" we may in the first instance be inclined to assume as self-evident. Indeed, the disadvantages accruing from great differentiations in adaptation to insect agency are really too important not to have been frequently noticed. Such are, "hercogamy," or the mechanical obstruction to self-fertilisation, as in Orchids; the physiological barrier, as in *Linum perenne;* the absence of insects required to fertilise a flower, as is the case with *Convolvulus sepium* in England, which rarely sets seed, as Sphinx Convolvuli is a rare insect; the frequent absence of bees, etc., in inclement weather, when Clover sets but little seed, to the great loss of the farmer; when certain flowers are neglected for greater attractions, as may be often seen when bees keep persistently to one species of plant and pass over others; the frequency with which bees perforate tubular flowers without pollinating them at all. Again, Muller points out* that while honey-seeking insects may legitimately cross heterostyled plants, pollen-seeking insects have no need to thrust their heads or proboscides down to the stigma of the short-styled forms; hence such tend to bring about illegitimate unions of the long-styled forms only. This, he thinks, may be a cause of the greater fertility of that kind of union.† Lastly, the more highly differentiated a flower is, the less is its number of insect visitors and the rarer may it become in nature. Thus orders of plants with easy access to the honey are some of the most abundant, as *Ranunculaceæ, Compositæ,*‡

* *Fertilisation*, etc., p. 387. † See above, p. 206.

‡ The enormous number of species and wide diffusion of the *Compositæ* are proofs of the advantages accruing to it from the peculiar

and *Umbelliferæ*; as well as are those dependent upon the wind, which néver fails, such as Willows, *Cyperaceæ*, and Grasses. On the other hand, all regularly self-fertilising plants are abundant, and, together with certain wind-fertilising plants, are cosmopolitan.

Although the idea that self-fertilisation is injurious is certainly not held now by botanists in so absolute a form as Mr. Darwin often stated it, yet it will not be amiss to point out the want of agreement between his conclusions and his own experiments.

In a chapter on "General Results,"* he commences by saying: "The first and most important of the conclusions which may be drawn from the observations given in this volume, is that cross-fertilisation is generally beneficial, and self-fertilisation injurious. This is shown by the difference in height, weight, constitutional vigour, and fertility of the offspring from crossed and self-fertilised flowers, and in structure of the flowers; first, in being adapted to a great variety of insects. Thus, on ten species of plants, Müller detected 546 species of insects, in the following proportions, Lepidoptera, 15 p.c.; Apidæ, 41 p.c.; Diptera, 27 p.c.; other short-tongued insects, 17 p.c. Bees, therefore, are the chief visitors. This is almost invariably the rule : the only species mentioned by Müller in his table in which short-lipped insects surpass in number the Apidæ is *Chrysanthemum leucanthemum*, which has a corolla tube, 3 mm. in length, in which the honey rises up into the widening throat and is easily accessible. The number of Lepidoptera is in the proportion of 6·9 p.c.; Apidæ, 16·6 p.c.; Diptera, 38·9 p.c.; others, 37·5 p.c. In *Achillea Millefolium*, with a corolla tube of 3 mm., Lepidoptera are 6·9 p.c.; Apidæ, 34·5 p.c.; Diptera, 24·1 p.c., and others, 34·5 p.c. Lastly, in *Centaurea Jacea*, with a tube of 7 to 10 mm., the Lepidoptera rise to 27 p.c.; Apidæ, 58·7 p.c.; while Diptera sink to 12·5 p.c., and other short-lipped insects are only 2 p.c.

The *Compositæ* thus well illustrate the fact that tubes are proportionate in length to the more specialized insects, a universal feature seen in all other orders as well.

* *Cross and Self Fertilisation of Plants*, p. 436.

the number of seeds produced by the parent plants. With respect to the second of these two propositions, namely, that self-fertilisation is generally injurious, we have abundant evidence. The structure of the flowers in such plants as *Lobelia ramosa, Digitalis purpurea*, etc., (1) renders the aid of insects almost indispensable for their fertilisation; and bearing in mind the prepotency of pollen from a distinct individual over that from the same individual, such plants will almost certainly have been crossed during many or all previous generations. So it must be, owing merely to the prepotency of foreign pollen, with cabbages and various other plants, the varieties of which almost invariably intercross when grown together. The same inference may be drawn still more surely with respect to those plants, such as *Reseda* (2), and *Eschscholtzia* (3), which are sterile with their own pollen, but fertile with that from any other individual. These several plants must therefore have been crossed during a long series of previous generations, and the artificial crosses in my experiments cannot have increased the vigour of the offspring beyond that of their progenitors. Therefore the difference between the self-fertilised and crossed plants raised by me cannot be attributed to the superiority of the crossed, but to the inferiority of the self-fertilised seedlings, due to the injurious effects of self-fertilisation."

Mr. Darwin then proceeds to discuss the first proposition, "that cross-fertilisation is generally beneficial," so that we may conclude that the preceding quotation represents the author's reasoning and conclusions on the idea of there being some "injuriousness" in self-fertilisation.

In the first place, it may be observed that the reason why Mr. Darwin's crossings yielded at first more marked results in height, fertility, etc., is because plants are never so carefully crossed in nature, nor self-fertilisation so carefully prevented, as was the case in his experiments. The probability is that the two processes are much more mixed in nature in the case of most plants. Therefore, by his experiments the more unalloyed influence of crossing brought about a much more enhanced stimulus than ever occurs in the wild state. Moreover, the prepotency of foreign pollen, upon which he lays stress, is a purely relative phenomenon; for whenever self-fertilisation yields more seeds than intercrossing, as is often the case, it is a just inference that the pollen "of the same flower" is then prepotent, in its turn. Indeed, Mr. Darwin actually found that in some cases intercrossing did no "good" at all, as in the case of the Garden Pea mentioned above, and in *Canna Warscewiczi*, etc.

I will now add some observations upon certain points I have numbered in this paragraph.

(1) That *Lobelia ramosa* and *Digitalis purpurea*, and many others given in a "List of Plants Sterile without Insect-aid," * cannot readily fertilise themselves unless the flower be disturbed in some way, is, *per se*, no proof that self-fertilisation is injurious; for the flowers of many of such plants are fully self-fertile when artificially assisted. Thus, Mr. Darwin says that although *Lupinus luteus* and *L. pilosus* seed freely when insects are excluded; yet Mr. Swale, of Christchurch in New Zealand, found Lupins only formed pods of seed when the stamens were artificially released, as they are not there visited at all by bees.† The interpretation of this fact, so well known that the term "hercogamous" ‡ has been invented for it, I take to be an immediate result of

* *Cross and Self Fertilisation*, etc., p. 357.

† *L.c.*, p. 150, note.

‡ If I remember rightly, by Errera; see *Bull. de la Soc. Bot. de Belg.*, xvii. (1887). The term means a "fenced-off union."

the action of insects. I have given reasons for believing, and the reader can readily suggest other instances, that structural peculiarities have grown in response to pressures and thrusts made upon the floral organs by the insects themselves; and that such have sometimes produced protuberances or obstructions in the way of the emission of the pollen upon the stigma of the same flower, is no more than might be anticipated to be extremely probable. Thus one of the most remarkable is the rostellum of Orchids, believed to be a modified stigma now converted to a new use. In nearly all Orchids this blocks up the way of access to the stigmatic chamber, while the pollen masses recline on the roof over it, so to say; but when Orchids become self-fertilising or even cleistogamous as well, this is often brought about by the degradation of the rostellum; so that the pollen masses can then easily slide over the summit of the stigmatic chamber and fall into it at once. When they do so they are fully self-fertile, as Mr. Henry O. Forbes has shown.*

Some few plants are quite barren with their own pollen, even when artificially placed upon the stigma; though *Lobelia* and *Digitalis* do not belong to the group. These, as shown elsewhere, can and often do become fully fertile at other places and seasons, and are thereby benefited by acquiring the possibility of setting seed by self-fertilisation, as otherwise they might set none at all.

There are, then, three kinds of barriers to self-fertilisation: one mechanical, as in Orchids; a second, that of time, when *e.g.* a flower is so strongly protandrous that the pollen is all shed before the stigmas are mature; and, thirdly, a physiological one, when the pollen is actually impotent on the stigma of the same flower, even though it be homogamous.

* *On the Contrivances for insuring Self-fertilisation in some Tropical Orchids*, Journ. Linn. Soc., xxi., p. 538.

THE VARIETIES OF FERTILISATION. 319

In no case is it logical to say that such arrangements are to prevent self-fertilisation. We may well ask why are a comparatively few plants thus provided for, and yet the vast majority are not. If, however, we regard them as results of differentiation brought about by the stimulus of insect agency—so that in certain places hypertrophy has set in and rendered the flower hercogamous, in others the androecium is so stimulated and its development so hurried on that the flower becomes protandrous, or its pollen so highly differentiated as to become like that of a distinct species,—we have a reasonable interpretation for these phenomena. Moreover, not one of them is absolute or stable. Thus a hercogamous Orchid can become self-fertilising; *

* Since the above was in type, Mr. H. N. Ridley has read a paper, at a meeting of the Linnean Society (Feb. 16, 1888), on "The Self-fertilisation of Orchids," in which he arrives at the same conclusions as Mr. H. O. Forbes (see above, p. 253, note), finding that the process is effected in several ways, especially, perhaps, by the degeneration of the rostellum. Moreover, the Orchids which he discovered to be capable of fertilising themselves are not only the most numerous in individuals, but are also the most widely dispersed of the genera to which they respectively belong. He also corroborates Mr. Forbes's observations, that Orchids set but a small percentage of their fruit, although fully exposed to the visits of insects.

Mr. H. Veitch has also contributed a valuable paper on the "Hybridisation of Orchids," in which he appears to corroborate M. Guignard's observations in every particular (see above, Chap. XVIII.).

The reader will take note of the significance of the fact that when Mr. Darwin published his work on "The Fertilisation of Orchids," it was thought that no flowers could equal them in their remarkable adaptations for securing the benefits of intercrossing by insect agency, and in their methods of "preventing self-fertilisation." Yet, of all flowering plants, evidence now tends to show that they set the least amount of seed, even when fully exposed to insects; while the order has furnished materials for two important papers on the many forms and ways by which self-fertilisation is secured in different genera.

the strongly protandrous Carnation can be made to be highly self-fertile, as Mr. Darwin showed; and *Linum perenne* can have its pollen so modified as to set seed abundantly in the same flower, as occurred with Mr. Meehan in Philadephia, though it was physiologically impotent in England.

It is, in fact, so to say a mere accident that mechanical and physiological barriers exist at all; and it is only by experiment that one can discover whether a flower so conditioned may not be really capable of self-fertilisation all the time. Indeed, Mr. Darwin's experiments have abundantly shown that self-fertilising properties are quickly reacquired, whenever the process is persevered with. For example, *Eschscholtzia Californica* was "absolutely self-sterile" in Brazil. Mr. Darwin, however, by self-fertilising it in England, raised the fertility in two generations to nearly 87 p.c.

When he asserts that his artificial crossings *could not have increased the vigour* of the offspring, and therefore all differences must be attributed to the inferiority of the self-fertilised, this argument would apply to a certain number of his experiments in different degrees, viz., with plants normally self-sterile; but he ignores the fact that, as soon as he tried to raise a stock of self-fertilised plants, the latter steadily gained upon the offspring of the crossed, till they equalled or surpassed them, or else would have done so had the experiments been continued.

Thus, with regard to *Lobelia ramosa*, the ratio of heights of the "intercrossed" to the "self-fertilised" offspring of first generation was 100 : 82; and the proportion of seeds as 100 : 60. In the second year, those growing under what he had proved to be the most disadvantageous condition for self-fertilised seedlings, namely, being crowded, the ratio of the heights became as 100 : 88·3. The experiment, unfortunately, was not continued further.

THE VARIETIES OF FERTILISATION. 321

Comparing this plant with *L. fulgens*, which is also quite sterile without aid, and, according to Gärtner, is "quite sterile with pollen from the same plant, though this pollen is efficient on any other individual,"* Mr. Darwin succeeded in raising self-fertilised plants by keeping the pollen of a flower in paper till the stigmas were ready, as it is strongly protandrous. The heights of the offspring were as 100 : 127, and Mr. Darwin adds, " the self-fertilised plants [in two out of four pots] were in every respect very much finer than the crossed plants."

In the next generation he used pollen from a different flower on the same plant to represent self-fertilisation. In this case those "self-fertilised" were only 4 p.c. below the crossed, the ratio being as 100 : 96. The conclusion, then, is that self-fertilisation *pure* was the best; intercrossing distinct plants, less so; and crossing on the same plant, the least.

Dianthus, like *Lobelia fulgens*, is strongly protandrous; but in the third generation the proportional number of seeds per capsule was as 100 : 125. " This anomalous result is probably due to some of the fertilised plants having varied so as to mature their pollen and stigmas more nearly at the same time than is proper to the species" (p. 135). *Exactly so.*

The conclusion I would draw is, therefore, not that self-fertilisation is *per se* in any way injurious, but that flowers which are normally sterile, by having become so highly differentiated through insect stimulation, do not now spontaneously set seed; and self-fertilisation is not so efficient as crossing. As soon, however, as the former process is persevered with, signs are not wanting of nature's showing even an eager response to it, till the results are often far superior to those normally obtained by intercrossing.

* *Cross and Self Fertilisation*, etc., p. 179.

THE STRUCTURE OF FLOWERS.

If flowers, unlike the preceding, are normally very self-fertile, as *Ipomœa* and *Mimulus* proved to be, then it appears that intercrossing supplies a remarkable stimulus, and the intercrossed beat the self-fertilised for a time. Sooner or later, however, the effect of the stimulus gradually disappears, and self-fertilisation reasserts itself. Thus with *Ipomœa purpurea* Mr. Darwin raised crossed and self-fertilised plants for ten generations; and the heights of the latter were 24, 21, 32, 14, 25, 28, 19, 15, and 21 p.c.,* respectively, less than the crossed. Grouping these into threes, the ratios become 100 : 74·3 ; 100 : 77·6 ; 100 : 81·6. That is to say, the intercrossed were steadily declining; for if the self-fertilised be regarded as 100, then the ratios of these to the crossed appear as follows : 100 : 134; 100 : 129 ; 100 : 121 Similarly with regard to fertility, the ratio of that of the intercrossed plants to the self-fertilised was for the first and second generations as 100 : 93 ; for the third and fourth, as 100 : 94; for the fifth, as 100 : 106 ; and the eighth, as 100 : 113. Hence the self-fertilised were superior.

Mimulus luteus gave analogous results. The crossed plants (*i.e.* offspring of crossings) surpassed the self-fertilised until the fourth generation, when several plants of the latter assumed a taller character, with whiter blossoms. This self-fertilising form "increased in the later self-fertilised generations, owing to its great self-fertility, to the complete exclusion of the original kinds." † "It transmitted its character faithfully, and as the self-fertilised plants consisted exclusively of this variety, it was manifest that they would always exceed in height the crossed plants." ‡

---

* These numbers correspond to the first nine years. The tenth gives 46; but Mr. Darwin thinks this number to have been accidental (p. 41).
† *Cross and Self Fertilisation*, p. 67.
‡ *Ibid.*, p. 70.

THE VARIETIES OF FERTILISATION. 323

(2) With regard to *Reseda* and *Eschscholtzia*, his observations are also somewhat misleading. Mr. Darwin experimented with *R. lutea* and *R. odorata*. They are both very capricious. Of *R. lutea* some individuals were absolutely self-sterile, whether left to themselves or artificially pollinated, while a few produced self-fertilised capsules. Similarly with *R. odorata*, when protected by a net some plants were loaded with self-fertilised capsules, others produced a few, and others, again, not a single one. Müller,* however, found that "plants which are kept protected from insects, yielded capsules filled with good seed." The inference from this variability in the fertility of different individuals in the same year, is that it is an accidental peculiarity of some to be more or less self-fertile than others; and that it was due to varying degrees of nutrition affecting the essential organs. We know now that plants frequently vary in their degrees of fertility, both at different seasons of the year,† and in different years or localities, according to climate, conditions of soil, etc. In any case, the self-sterility of these plants is by no means so absolute as to justify the belief of their having never been self-fertilised for years.

Let us now turn to Mr. Darwin's experiments.

Reseda lutea. The ratio between the heights of the crossed plants and those of the self-fertilised were as 100 : 85, the weights as 100 : 21, when the plants were grown in pots. When grown in open ground they were nearer equality, viz., in height, as 100 : 82, and in weight (a better test than height), as 100 : 40. Differences in fertility are not given, and, therefore, presumably not striking.

* *Fertilisation*, etc., p. 116.
† Mr. Darwin says *Papaver vagum*, included in the list of plants sterile without insect aid, produced a few capsules in the early part of the summer; see above, Chap. XXV., on Sexuality and Environment.

Reseda odorata. The results of plants grown in pots were as follows, the proportions being taken as before. The heights were as 100 : 82; weights as 100 : 67; while their heights when the plants were grown in the open were as 100:105.*

He next raised seed by crossing some flowers and self-fertilising others on the same plant of a particular semi-self-sterile individual. From these the seedlings gave the following results: heights as 100 : 92; weights as 100 : 99; fertility as 100 : 100.

These results show that the differences have practically vanished; the weight being a much better test than height, as it points to greater assimilative powers, and leaves nothing to be desired.

It is difficult, then, to see how *Reseda* furnishes data for any argument raised to prove the existence of injuriousness in the self-fertilisation of plants. Indeed, Mr. Darwin himself observes: "I expected that the seedlings from this semi-self-sterile plant would have profited in a higher degree

* Mr. Darwin remarks upon this result as follows: "We have here the anomalous result of the self-fertilised plants being a little taller than the crossed, of which fact I can offer no explanation. It is, of course, possible, but not probable, that the labels may have been interchanged by accident" (*Cross*, etc., p. 121). In my paper (p. 383) referred to I have shown that it was most generally the case that while a close competition in the same pot proved disadvantageous to the self-fertilised seedlings, yet, when they had no competition, the differences were not nearly so marked. There are apparently but two alternatives to appeal to in order to account for the fact that intercrossed plants are not so greatly superior to the self-fertilised when planted in open ground, as when in competition in pots; viz., either the intercrossed plants become deteriorated on being planted in open ground, which is absurd, or else the self-fertilised must regain or acquire vigour in a *relatively greater degree* than do the intercrossed, and thus would seem to evince what might be called a greater "elasticity" of growth than their intercrossed competitors.

from a cross than did the seedlings from the fully self-fertile plants. But my anticipation was quite wrong, for they profited in a less degree : " *—really not at all, for the self-fertilised were superior. "An analogous result followed in the case of *Eschscholtzia*, in which the offspring of the plants of Brazilian parentage (which were partially [said to be "absolutely" so, on p. 111] self-sterile) did not profit more from a cross, than did the plants of the far more self-fertile English stock." *

Mr. Darwin commenced his experiments by saying, "This plant is remarkable from the crossed seedlings not exceeding in height or vigour the self-fertilised. On the other hand, a cross greatly increases the productiveness of the flowers on the parent-plant, and is sometimes necessary in order that they should produce any seed. Moreover, plants thus derived are themselves much more fertile than those raised from self-fertilised flowers ; so that the whole advantage of a cross is confined to the reproductive system." †

Twelve flowers crossed produced eleven good capsules, containing 17·4 grains of seeds; eighteen self-fertilised flowers produced twelve good capsules, containing 13·61 grains : therefore the ratio of fertility was as 100 : 71. In the first season the heights were as 100 : 86. Being cut down, the next season, they were reversed, "as the self-fertilised plants in three out of four pots were now taller than and flowered before the crossed plants."

"In the second generation, eleven pairs were raised and grown in competition in the usual manner. The two lots were nearly equal during their whole growth, or as 100:101. There was no great difference in the number of flowers and capsules produced by the two lots, when both were left freely exposed to the visits of insects."

* *Cross and Self Fertilisation*, p. 121.   † *L.c.*, p. 109.

THE STRUCTURE OF FLOWERS.

This concludes his experiments with English plants; and though crossing did little or no good, and the first average of heights, viz. 100 : 82, he thinks were accidental, the converse proposition, that self-fertilisation was injurious, is in no way proved. It would be just as logical to say that, since the self-fertilised plants grew more vigorously after both were cut down, that crossing must have weakened the constitution of the crossed seedlings. Or, again, from the second year's results, we might justly conclude that the two effects were quite identical.

He next experimented with seed the parents of which had been cultivated in Brazil, in which country Fritz Müller had found them to be "absolutely self-sterile with pollen from the same plant, but perfectly fertile when fertilised with pollen from any other plant." Seeds raised from these in England "were found not to be so completely self-sterile as in Brazil." The average number of seeds produced in the capsules borne on the intercrossed and self-fertilised plants of Brazilian origin were 80 and 12 respectively in the first year; that is in the ratio of 100 : 15.

With regard to the second generation, or grandchildren, next raised, Mr. Darwin observes : " As the grandparents in Brazil absolutely required cross-fertilisation in order to yield any seeds, I expected that self-fertilisation would have proved very injurious to these seedlings, and that the crossed ones would have been greatly superior in height and vigour to those raised from the self-fertilised flowers. But the result showed that my anticipation was erroneous ; for as in the last experiment with plants of the English stock, so in the present one, the self-fertilised plants exceeded the crossed by a little in height, viz., as 100 : 101."

In the next year the average number of seeds per capsule of the crossed and self-fertilised was as 100 : 86·6 ; so that the relative fertility of the self-fertilised had risen from zero in Brazil to 15, and then to 86·6 p c., in comparison with the crossed regarded as 100.

He now made crossings between the offspring of the Brazilian plants and the English-grown plants, with the following results:—

First, as to heights,—

The English-crossed to the self-fertilised plants	... ...	100 : 109
The English-crossed to the intercrossed * plants	... ...	100 : 94
The intercrossed to the self-fertilised plants ...	... ...	100 : 116

Secondly, as to weights,—

The English-crossed to the self-fertilised plants	... ...	100 : 118
The English-crossed to the intercrossed plants	... ...	100 : 100
The intercrossed to the self-fertilised plants ...	... ...	100 : 118

Three rows of plants of each kind grew in the open; and here also the self-fertilised grew taller than the others. Moreover, *all* except three of the self-fertilised were killed by the winter.

"We thus see that the self-fertilised plants which were grown in the nine pots were superior in height (as 116 : 100) and in weight (as 118 : 100), and apparently in hardiness, to the intercrossed plants derived from a cross between the grandchildren of the Brazilian stock. The superiority is here much more strongly marked than in the second trial with the plants of the English stock, in which the self-fertilised were to the crossed in height as 101 : 100. It is a far more remarkable fact . . . that the self-fertilised plants exceeded in height (as 109 : 100), and in weight (as 118 : 100), the offspring of the Brazilian stock crossed by the English stock."

* "Intercrossed" signifies the offspring of the Brazilian plants crossed with one another.

When we look back and remember that the plant was "absolutely self-sterile" in Brazil, and compare that fact with these final results, it is difficult to see how self-fertilisation can be charged in any way with injuriousness. Though the results may have shown little or no advantage from crossing, it does not follow "that the differences," namely greater height, weight, or fertility of the self-fertilised, were attributable "to the inferiority of the self-fertilised seedlings, due to the injurious effects of self-fertilisation."

On the other hand, the facts appear to warrant the conclusion that this north-temperate plant became barren in Brazil in consequence of the hot climate; that the recovery of its self-fertilising powers was due to the English climate better suiting it; that it at once responded to the effort, so that its self-fertility rose in two generations from 0 to 86·6 p.c. The plants, too, thus raised showed nothing to indicate any constitutional derangement that might, with any show of reason, be attributable to self-fertilisation.

From the preceding observations upon Mr. Darwin's reasoning, I think the reader will now see that it is not so conclusive in proving the existence of any injuriousness in self-fertilisation as he appeared to think.

This chapter was already in type when I met with the following passage in "The Life and Letters of C. Darwin," written in May, 1881 : "I now believe . . . that I ought to have insisted more strongly than I did on the many adaptations for self-fertilisation, though I was well aware of many such adaptations."

With regard to the values of other kinds of fertilisation, I must refer to Mr. Darwin's works; for it is beyond my purpose to discuss them, as they have no special bearing upon the origin of floral structures.

CHAPTER XXXII.

FERTILISATION AND THE ORIGIN OF SPECIES.

THE ORIGIN OF SPECIES BY INSECT AGENCY.—The attractive features of flowers being now well recognized as correlated with insect agency in fertilisation, the question arises, How have they come into existence? We may suppose that a plant bore seedlings, some of which had, we will say, the corolla accidentally (that means from some unknown cause arising from *within*) larger on one side than another; and then such a flower, being selected by insects, left offspring which, by gradual improvement through repeated selection, ultimately reached the form it now possesses.

As an alternative, we may suppose that the first impulse came from *without*, and induced by the insect itself; so that the variation once set up in a definite direction, went on improving under the constantly repeated stimulus of insect visitors until the form of the flower was actually conformable to the insect itself.

The process of evolutionary development might perhaps be much the same under either supposition, but the latter hypothesis has more than one advantage. First, in the assignment of *a direct physical cause* for the incipient change, instead of some incidental and unaccountable variation, which must be assumed by the former. Secondly, the theory does not require the plant to make an indefinite number of less useful changes or variations, only to be discarded at each generation for the one form that was wanted. Thirdly, as a great number of flowers would be visited, both on one plant and on many surrounding individuals in the neighbourhood, great numbers might bear offspring advancing more or less in the same direction; and there would be no fear of extermination, even if some happened to be crossed by the parent form. Indeed, the varying offspring would largely supersede the parent form in number altogether, if they sprang up at one place without emigration. If we supply the additional aid of isolation, many other influences would be brought to bear upon them, and they would be free to vary without any interference from the parent stock.

Mr. Darwin has abundantly shown that when a plant is crossed, and its seedlings struggle in a confined place with those derived from flowers which have not been crossed but artificially self-fertilised, they generally succeed in mastering the latter; so that if there be any struggle with the seedlings of a self-fertilised parent, such a struggle for life is mainly during the early period of growth, before any varietal or specific characters of the flowers have put in an appearance at all. For it is only in the youthful stages that the greatest contest is maintained; and the result depends largely upon *constitutional*, and not at all upon *specific*, that is morphological characters, mostly taken from the flowers. Now, Mr. Darwin has shown that such constitutional vigour does very generally accompany at least the first few years of crossing. So that we have a *vera causa* of the success of such newly crossed plants in the preliminary struggle for life. It need hardly be remarked that if insects thus start a new variety, they are crossing the flowers at the same time.

It is true that the stimulus of crossing does not last for

many years; but it is probably all that is wanted to give the crossed plant the ascendancy when starting on an evolutionary career.

As an illustrative case of a struggle between two varieties, I took the same quantity of English-grown "Revett's" wheat and Russian "Kubanka," the former having a preponderance of starch and the latter of gluten, being a smaller and harder grain. I sowed them as thickly as possible on a square yard, the two kinds having been previously well mixed together. They all germinated, and the struggle of course became intense. About twenty ears only were produced, which were all Kubanka. The experiment was repeated a second year, with the same result. This is what I would consider as, therefore, due to "constitutional selection."

Survivors, however, are by no means entirely dependent upon constitution, much less on specific differences; for seeds which fall on the circumference of the crowd, or on a better soil than that upon which others may happen to lie, as on stony ground, are thereby "selected," but it is through no merit of their own, as in any way being the fittest, for they survive only because they are the "luckiest;" just as out of the thousands of eggs of a salmon a few only escape the jaws of their enemies: so that simply "good luck" plays an important part in determining which shall survive and come to maturity in both kingdoms alike.

Hence, during the period of life when the struggle for existence is most intense, there are various circumstances which determine what plants shall survive; and in probably few cases, generally no case, have the morphological variations or specific characters any voice in the matter of selection whatever, excepting indirectly, as stated above, whenever constitutional vigour is correlated with first crossings.

The difficulty which Mr. Romanes has felt in the struggle for life through the swamping effect of a varying offspring being crossed with the parent form, seems to me to be illusory as far as most flowering plants are concerned.* For not only do the majority of new forms arise through transport of seeds to a new and distant locality, but even at home, if the plant be at all responsive, so many seedlings, perhaps all, will tend to be differentiated at the same time and in the same way, that the parent form will soon be in a minority, and if now neglected by insects may die out through "insect-selection" of the new form.

According to the old view, that plants are varying spontaneously in all directions, and that only a few are selected by insects, the difficulty has long been felt that dangers of all sorts must surround the offspring of those few. Let us reverse the process, however, and let the insects themselves be the cause of changes set up in the flowers in the adaptive directions, and the responsive power of the flower itself will soon develop the best forms. These run no risk of being lost, through the multitude of offspring. Hence, if my theory be true, physiological selection, which I cannot find horticulturists are inclined to accept, is not needed at all.

Suppose some prevailing insect to have begun to set up incipient changes for a new variety, which then becomes dispersed; since many of the offspring will possess the new adaptation, and several other kinds of insects will visit the flower in different places, as the seeds happen to get transported, the result will be, that while the original species of insect induces the descendants of the plant at home to vary in adaptation to itself, others are at work elsewhere,

* Fritz Müller found the genus *Abutilon* and a species of *Bignonia* to be more or less sterile with parental pollen. See Müller's *Fertilisation*, etc., pp. 145, 466.

modifying the same incipient alterations to suit themselves. Hence, as soon as isolation by migration has taken place, it is the presence of other insects which determines the development of other varieties. All, however, are based on the same plan of departure.

In this way many varietal and subsequently specific forms of the same genus will arise; and the further they travel from the parental home the greater, perhaps, will be the specific differences; and thus can representative species be accounted for, especially among conspicuously flowering plants.

On the other hand, the perpetually self-fertilising species which alone, as a rule, are cosmopolitan, are almost identical in form, or at least have a minimum of differences between them, and such as may possibly be accounted for by climatal causes alone.

DIFFICULTIES OF NATURAL SELECTION.—The greatest difficulty I have always felt in the idea that a plant was selected because it had some floral structures more appropriate than others, lay first in the fact that the principal period of the struggle for life takes place in the seedling stage, before any varietal and specific characters have appeared; and, unless there were a large number of the seedlings which would ultimately bear the improved flower, or else a superior constitutional vigour be guaranteed to be correlated with the particular varietal characters to be preserved, these alone could have nothing to do with the survival of the fittest.

Secondly, granting that the plant has succeeded in surviving till the flowering period, then why should so many minute details of floral structure be *necessarily* correlated? If the loss of three out of five carpels in the *Labiatæ* were due to natural selection, why should this go hand-in-hand with a multiplication of the ribs of the calyx, and the peculiar lipped and hooded corolla with the lateral position of the flower, etc.? We find in selecting peas and beans great varieties among them, but next to none in the calyx and corolla, to which the horticulturist pays no attention.

In nature, however, we often find in flowers regularly visited by insects innumerable and minutely correlated adaptations in all the whorls, which must have all varied together to form such existing flowers. Now, the difficulty of their doing so without some common cause, which affects them all simultaneously, seems to me insuperable.

If my theory, however, be accepted, it solves the whole mystery at once, as all the changes are set up by one prime cause, namely, the irritations of the insect in the case of all flowers adapted to insect-fertilisation; while the absence of insects in regularly self-fertilised flowers, as well as anemophilous ones, is sufficient to account for the atrophy which has affected them, the present condition of such flowers having been the inevitable result.

Hence, instead of speaking of the Origin of Species of Plants by Natural Selection, I would regard the survival of the fittest as first issuing from "Constitutional Selection;"* while the origin of the floral specific characters is the result of the responsive power of protoplasm to external stimuli. These latter are infinitely various in kind and degree, as has been shown in the early part of this book. The result is, that while high differentiations occur in some directions, degradations are met with in others, sometimes seen in different parts of the entire plant; but not at all infrequently are both features observable simultaneously in one and the same floral

* Of course the chances of less competition by growing on the circumference of the batch of seedlings, by receiving a little more light, etc., aid in selecting, and sometimes may determine, as stated above, those which shall survive.

whorl. The phrase "natural selection" will therefore have been noticed as conspicuous by its absence throughout this book. This is not because I would in the least deny the fact that vast numbers of seedlings perish while others survive through that form which I have called "constitutional selection," which are thus "selected," and arrive at the flowering and fruiting stages; and, again, that of these latter many may set no seed through the neglect of insects, etc., and so perish entirely and leave no offspring, while others again survive and are selected. Why, however, I do not refer any particular structure to the action of natural selection is because I have always felt or perceived a danger in doing so. Natural selection is, as thus styled, an *abstraction;* and as long as we hide our ignorance of its *concrete* representatives, that is to say, the real causes at work to induce a change, we may fancy we understand all about it, while we may be in reality in profound ignorance.

Professor Huxley remarked, in his lectures on the *Origin of Species,* that what we want is "a good theory of variation." It is in the attempt to fill this hiatus that I have, step by step throughout this book, preferred to give what seemed to me a direct cause, mechanical, physiological, climatal, etc., for every structure; which may bring us nearer to a comprehension of the direct interaction of cause and effect than the vague term "natural selection" seems capable of doing Thus, to take an example, Müller refers the loss of the fifth stamen in Labiates to natural selection, but makes no statement *how* he supposes selection to have done it. On the other hand, I would prefer to attribute its absence to atrophy, in compensation with the hypertrophy of the corolla on the posterior side. I may be wrong, of course, but at all events I give a reasonable cause, which is a fertile one in bringing about alterations in the structure of flowers; whereas "natural selection" leaves us exactly where we were before. Moreover, natural selection is made to cover exactly opposite processes; for the formation of the enlarged lip, on the one hand, would be attributed to it, just as much as the elimination of a stamen altogether, on the other. Instead, therefore, of using this term as the cause of anything and everything, I prefer to attribute effects to hypertrophy, atrophy, resistance to strains, responsive action to irritations, and so on. If it be thought that natural selection somehow underlies all this, the reader is at liberty to substitute the phrase; but, I must confess, it conveys nothing definite to my mind, while the others undoubtedly do.

I do not wish the reader to suppose that my theory is altogether in opposition to Mr. Darwin's; for it must not be forgotten that he himself laid great stress on the environment as a cause of variability upon which, when once brought about, natural selection could then act. Thus he remarks: "To sum up on the origin of our domestic races of animals and plants. Changed conditions of life are of the highest importance in causing variability, both by acting directly on the organisation, and indirectly by affecting the reproductive system. It is not probable that variability is an inherent and necessary contingent, under all circumstances. . . . Variability is governed by many unknown laws, of which correlated growth is probably the most important. Something, but how much we do not know, may be attributed to the definite action of the conditions of life. [Under this I would include the definite action of insects exerted mechanically upon the organs of flowers.] Some, perhaps a great, effect may be attributed to the increased use or disuse of parts. [Compensation plays undoubtedly a very important part]. . . Over all these causes of Change, the accumulative action of Selection, whether applied methodically and quickly, or

FERTILISATION AND THE ORIGIN OF SPECIES. 337

unconsciously and slowly, but more efficiently, seems to have been the predominant Power." *

If thus the variations of floral structures can be reasonably referred *directly* to external agencies, and we may speak of each as a cause instead of using the abstract expresssion "natural selection," there still remains the question, What has brought into existence the primary flowers themselves, which insects have subsequently modified into their present conditions ?

THE ORIGIN OF FLOWERS.—There are good reasons for regarding Gymnosperms—both from their extreme antiquity, as well as from points of structure showing affinity with the higher Cryptogams; such, for example, as the *Lycopodiaceæ*—as standing in some sort of way intermediate between the latter and Dicotyledons. Yet the connecting links are much wanted on both sides of them. As far as *Coniferæ* and *Cycadeæ* can help us, we are strongly led to believe that they were primitively, just as they are now, anemophilous and diclinous; though the subdiœcious (?) *Welwitschia* has points of structure which seem to indicate its being a degraded state of an hermaphrodite plant. This remarkable monotypic genus is, however, too isolated and unique to afford any safe point of departure on the road to Dicotyledons, so that with regard to the latter we are still driven to speculation alone.

If, then, we are right in assuming Gymnosperms to have been always diclinous, and Dicotyledons to have arisen from some member of that group, then it is presumable that the first were diclinous, perhaps diœcious, and anemophilous as well. The general opinion seems to be that they were diœcious; and Mr. Darwin thought that monœcism was the next step, and thence hermaphroditism was ultimately reached.

* *Origin of Species*, 6th ed., p. 31.

Now, we must not forget that when a female flower is pollinated the effect of the impregnation by the pollen-tube is not only to create an embryo in the ovule, but to endow it *potentially* with its own sexuality; so that the sexless embryo becomes potentially both male and female; in as much as it may subsequently grow up to be solely a male or solely a female plant; or else it may combine the sexes, either as a monœcious or hermaphrodite plant.

Moreover, we now know that the resulting sex which appears in diœcious plants on maturity is largely, if not entirely, dependent upon conditions of nutrition, possibly aided by other and unknown influences.

Consequently, we cannot say for certain whether the first Dicotyledons were not at least monœcious, if not hermaphrodite, since the former of these states prevails already in Gymnosperms, as in *Pinus;* while the latter is hinted at in not infrequent monstrous conditions when the lowermost scales of the spiral series in cones of *Abies excelsa*, etc., are antheriferous, instead of being ovuliferous.* Such cases show that one (the male) sex can *suddenly* appear in the same spiral series as the other. And this is all that is wanted to form an hermaphrodite flower; for continuously spirally-arranged sexual organs are characteristic of many plants, such as of the *Ranunculaceæ*; and such a monstrous condition *may* simply be a reversion to a primitive hermaphrodite state. Hence appears the inherent possibility of the production of hermaphroditism without any slow evolutionary process at all; but simply as a result of the conveyance of the male energy to the female plant, by the very act of pollination itself.

Mr. Darwin, when speculating on the origin of hermaphroditism, wrote as follows: " By what graduated steps

* *Teratology*, p. 192.

an hermaphrodite condition was acquired we do not know. But we can see that if a lowly organised form, in which the two sexes were represented by somewhat different individuals, were to increase by budding either before or after conjugation, the two incipient sexes would be capable of appearing by buds on the same stock, as occasionally occurs with various characters at the present day. The organism would then be in a monœcious condition, and this is probably the first step towards hermaphroditism; for if very simple male and female flowers on the same stock, each consisting of a single stamen or pistil, were brought close together and surrounded by a common envelope, in nearly the same manner as with the florets of the *Compositæ*, we should have a hermaphrodite flower." *

It is a singular fact that Mr. Darwin never seems to have thought of *Euphorbia*, which tallies exactly with his hypothetical origin of a hermaphrodite flower; but, unfortunately, a "blossom" of an *Euphorbia* is *not* regarded by botanists as a flower, but an inflorescence. It consists of a "single pistil," on its own pedicel, surrounded by many "single stamens," each on their own pedicels; and are "brought close together and surrounded by a common envelope."

Mr. Darwin's mistake resides in his supposition that hermaphroditism must have arisen from diœcism, by passing through monœcism; so that he is obliged by this order of progress to consider a flower with stamens and a pistil to be made of separate flower-buds, *i.e.* to be *axial structures* with their appendages reduced to at least one of each kind. But from phyllotactical reasons, it is clear that the origin and arrangements of the floral members are entirely foliar.

All that seems necessary for us to assume as the origin of a flower with a conspicuous corolla or perianth, is a leaf-bud

* *Cross and Self Fertilisation of Plants*, p. 410.

of which some of the members have already differentiated into carpellary, others into staminal organs, the outer appendages being simply bracts, like, we will say, those surrounding the stamens or ovule of the Yew.

As insects often come for pollen alone—as in honeyless flowers of Laburnum, Poppies, St. John's Wort, and Roses,— and then pierce the juicy tissues for moistening the honey, as they have been seen to do in *Anemone*, Laburnum, Hyacinths, *Orchis*, etc., we may, I think, infer with some probability that they did the same with the primitive flowers.

Having once attracted insects to come regularly, then a multitudinous series of differentiations would follow. The corolla would in all probability be the first to issue out of the bracts, as being the next whorl to the stamens and as a result of stimulus; other changes, already described under the Principles of Variation, would follow by degrees and in different combinations; but in every case they would be due to the responsive action of the protoplasm in consequence of the irritations set up by the weights, pressures, thrusts, tensions, etc., of the insect visitors.

Thus, then, do I believe that the whole Floral World has arisen.

INDEX.

A

Adelphous filaments, 57; imitated, 59; and nectaries, 58
Adhesion, analogies in animal kingdom of, 48, 88; principle of, 5, 78, seqq.; rationale of, 80; of stamen to perianth, and origin of, 81, and to style (?), *Aristolochia*, (fig. 21) 83
Æstivations and phyllotaxis, (fig. 3) 15
Alpine, flowers, colours of, 176; strawberry, phyllody of, 301
Amaryllis, appendage to perianth, (fig. 41) 134
Androdiœcism, examples, explanation and origin of, 227
Andrœcium, explained, 4; irregularity in, origin of, 109
Anemophilous flowers, 265, seqq.; characters of, 268; cosmopolitan, 283; "long-lived" stigmas of, 269; pollen of, 266
Anemophily, and Greenland flora, 270; and cleistogamy, 264; and degeneracy, 266, 272; and heterogamy, 269; origin of, 266, 270, 272; and protogyny, 200, 269, 272
Anisomerous whorls, explained, 5; causes of disarrangement of, 45
Anthers, on bracts, (fig. 64) 288; connivent, of Violet, 60; contabescent, 275; on glumes, (fig. 65) 288; metamorphosed, 293, (fig. 81) 298, (figs. 83, 84) 302; stigmatiferous, (fig. 76) 294; syngenesious, and interpretation of, (fig. 11) 60; versatile, 266, 268
Ant-plants, hereditary effects of irritation in, 115, 142, 157
Appendages, in *Amaryllis*, (fig. 41) 134; and axis, homology between, 309; origin of floral, 133
Aquilegia vulgaris, arrangement of floral whorls of, 22; number of parts in, 22
Arabis albida, leaf-traces of, (fig. 7) 39
Arctic flora, and anemophily, 270; and self-fertilisation, 259
Aristolochia, structure of flower, (fig. 21) 83
Arrangement, causes of, 47; displacement of, by anisomery, and substitution, 45; illustrations of, in *Ranunculaceæ*, 21, seqq.; principle of, 5, 139
Arrest, of carpels, 4, 8, 278; of carpels in *Campanulaceæ*, 44; of floral axis, 6; in free-central placentas, 72, seqq.; of growth of ovary and seeds in Orchids, 169, 281, and in Willows, 170
Atragene, staminal nectaries of, (fig. 44) 141
Atrophy and hypertrophy in animal kingdom, 88; as causes of irregularities, 108; in compensation, 105; in zygomorphism, 116, seqq.

INDEX.

Autogamy, explained, 198, 311. *See* Self-fertilisation.
Axis, and appendage, homology between, 309; floral, cause of arrest of, 6

B

Beta, formation of ovule of, (fig. 16) 73
Boughs, curvature of, due to strain, (fig. 39) 125
Bracts, petaloid, 286, (figs. 62, 63) 287; pistiloid (glumes), (fig. 65) 288; progressive changes in, 286; transitional forms of, in Hellebore, (fig. 61) 286
Bulbs, origin of, from funicle, 310; from leaf-sheath, 310

C

Cabbages, excrescences on, homologous with ovules, 307
Calyx, arrest of, 8, 184, 194; progressive metamorphosis of, 288; -tube, 89, seqq. *See* Sepals.
Campanula medium, anatomy of flower of, (fig. 8) 43, (fig. 15) 71
Campanulaceæ, arrangement of carpels in genera of, 44
Capparideæ, andrœcium of, and symmetry in flower of, 33
Carpels, arrest of, 4, 8, 278; in *Campanulaceæ*, 44; cohesion of, 62; decrease by compensation, 21, 278; phyllody of, 302; superposition of, 44, seqq.; typical number of whorls of, 4. *See* Pistil.
Carpophore, placental origin of, 72
Cell-division and light, 154
Cell-wall, thickening of, to resist pressure, 127
Centaurea, adaptations for fertilisation, (fig. 11) 60; and sexuality, 240
Change of symmetry, 18, 186
Chorisis, and arrangement, 24, 39, 44,

46; multiplication of stamens by, 44, and of carpels by, 44, 308, and of ovules by (in Orchids), 309
Cleistogamy, and anemophily, 264; and degeneracy, 251, seqq.; and environment, 263; explained, 198; in flowers, 251; illustrations of, 257-262; in *Impatiens*, (fig. 58) 261; in *Lamium*, (fig. 59) 261; origin of, 262-264; in *Oxalis*, (fig. 57) 260; in *Salvia*, (fig. 60) 262; in Violets, (figs. 55, 56) 257, 258
Cohesion, of carpels, 62; illustrations of, 49, 50; origin of, 50; of petals, 56, in *Phyteuma*, (fig. 9) 50; principle of, 5, 48; of sepals, 54, of stamens, 57, to resist strains, 51, 53; varieties of, congenital and by contact, 48
Colours, of Alpine flowers, 176; changes in, 176; and darkness, 177; effect of crossing on, 178; effect of salts on, 175; of flowers, 174; and insects, 182; laws of, 174; nutrition and, 178; origin of, 178; as pathfinders, 178, and arrest of, 253, white and pale tints, and self-fertilisation, 253; whole, and self-fertilisation, 183
Compensation, in adaptations of flowers, 105, 117; atrophy and hypertrophy in, 105; increase of seeds and decrease of carpels by, 21, 278; in irregular flowers, 103, seqq.; in rudimentary organs, 284
Conducting tissue, of Orchids, 165; origin of, by irritation of pollentube, 165, seqq.; structure of, (fig. 50) 164
Coniferæ, foliage of, adnate and free, 84; origin of flowers and the, 337
Connivent anthers, of Violet, 60
Contabescence of anthers, 275
Cords, fibro-vascular, alteration in orientation of, 64, 65; as floral units, 300, 308, 309; in flower of *Campanula*, (fig. 8) 43, (fig. 15) 71; increase in number of, 55-57; orientation of phloëm and tracheæ

INDEX. 343

in, 63; in receptacular tubes, (fig. 14) 68, (fig. 28) 95, (fig. 30) 97; sepaline, of *Salvia*, 55; as origin of the staminal and carpellary, in *Malvaceæ*, 43, 44
Corollas, appendages to, origin of, 133, seqq.; form of, 101, seqq.; metamorphoses of, 292, 301; movements in, of *Genista*, (fig. 47) 160; of *Lopezia*, (fig. 48) 161; origin of, irregular, 103, seqq.; petals of, displacement of, by insects, (figs. 33-35) 110, 111; polliniferous, 292, 293; progressive metamorphoses of, 292; reduction of size of, 9, 254, in *Geranium*, 252; regular and irregular, 101, seqq.; sensitiveness in, *Ypomœa*, 161; stameniferous, (figs. 72, 73) 292, 293; strains, effect of, on the formation of, 101, seqq., 126; structure of bilateral, 116, seqq.; virescence of, (figs. 83, 84) 301, seqq. *See* Petals.
Correlation of growth, 112, 113, 117; irregularities by, 108
Cross-fertilisation, advantages of, in evolution of species, 330, and in horticulture, 311; colour, effects on, 178; disadvantages of, 314; rationale of, 312; stimulus produced by, 312; views of Mr. Darwin on, 315
Cruciferæ, anatomy of floral receptacle, (fig. 6) 32; symmetry of, 32

D

Darkness and colours, 177
Declinate stamens, in *Dictamnus*, (fig. 33) 110; distribution of forces in, of *Echium*, (fig. 20) 82; of *Epilobium*, (fig. 34) 111; origin of, due to weight of insects, 110, 111
Degeneracy and degradation, of andrœcium, 273; and androdiœcism, 227; and anemophily, 266; of flowers, 251, seqq.; in inconspicuous flowers, cause of, 251; in Orchids,

172, 281, 319; origin of, 282; and self-fertilisation, 252, seqq.
Development, of floral whorls, 191, and continuous during flowering, 122; order of, of parts of flowers, relative only, 195; rates of, in pistil, 192, 193
Dialysis, explained, 5, 50; in *Mimulus*, (fig. 10) 51
Diclinism, and heterostylism, 228; partial, 220; in primitive flowers, 337
Dimorphism, and fertilisation in *Viola tricolor*, 255; and heterostylism, 203; in stamens, (fig. 37) 121
Diœcism, and heterostylism as cause of, 218; of primitive flowers, 337
Domatia, hereditary formation of, 115, 142, 157
Doubling, causes of, 298
Drosera, metamorphoses of tentacles of, into ovules, 307
Duvernoia, zygomorphism of, origin of, (fig. 31) 107

E

Electricity, effects on protoplasm, (fig. 45) 152, on nucleus, 154
Emergence, alteration in order of, in regular and in irregular flowers, 187; and development of ovules, 195, and interpretation of, 196; of floral whorls, 184; order of, 184
Energy, reproductive and vegetative, 231, seqq.
Environment, action of, Mr. Darwin's views on, 336; influence of, 158; origin of species through, 329, seqq. *See* Preface.
Epidermis, origin of root hairs on, (fig. 42), 137
Eranthis, arrangement and number of parts in flower of, 22
Exclusion, of insects from flowers, 102, 133, seqq.
Excrescences, on corolla, (fig. 87) 306; on cabbage-leaves, as homologues of ovules, 307

F

Fasciation, 51, 85; of petioles of pear, (fig. 26) 94
Fertilisation, cross- (see s.v.); and origin of species, 329; by pollen-tube (see s.v.); varieties of, 311; self- (see s.v.)
Fibro-vascular cord, as a fundamental unit, 300, 308, 309. See Cord.
Flora, of Dorrefjeld, and self-fertilisation, 259; of Galapagos Islands, 270; of Greenland, 270
Floral symmetry, correlation with phyllotaxis, 14; explained, 4, 5; variations in, 12
Floral whorls, development of, order of, 191; emergence of, 184; symmetrical decrease and increase in, 18; unsymmetrical, 20. See Whorls.
Flowers, conspicuous, development of parts of, 191; degeneracy in, 251; inconspicuous, origin of, 251; origin of, 337; typical, structure of, (fig. 1) 3
Forces, effects of mechanical, etc. See Mechanical forces.
Forms, of floral organs, 101, seqq.; dimorphic, of stamens, (fig. 37) 121; principle of, 5; transitional, 118, seqq.
Funicle, bulb arising from, 310; as origin of ovule, 303

G

Galls, analogous to tumours, 144; due to irritation, 144; hairs of, 138
Garidella, arrangement and number of parts in, (fig. 4) 21
Glands and rudimentary organs, 283. See Nectaries.
Growth of organs, continuous during flowering period, 122; correlation of, 112, 333
Guides, degeneracy of, in self-fertilised flowers, 253; origin of, 178

Gymnosperms, and the origin of flowers, 337
Gynandrous, 82; *Aristolochia*, (fig. 21) 83
Gynodiœcism, causes of, 221, seqq.; and climate, 221; explained, 220; origin of, 222, seqq.; and soil, 221
Gynœcium, degeneracy of, 278; explained, 4; unsymmetrical decrease in, 20. See Carpels and Pistil.
Gynomonœcism, examples of, 226; explained, 220

H

Hairs, on filaments, origin of, 136 (see fig. 11, 60); in galls, 138; on roots, origin of, 137; on seeds, 170; within styles, origin of, 139; tangles and wheels, origin of, 133, seqq.
Heliotrope, stigma of, cause of anomalous, 135
Hellebore, alteration in orientation of cords, (fig. 12) 64; arrangement and number of parts of floral whorls (fig. 5), 22
Hercogamy, explained, 317; in Orchids, 314; relative character of, 319
Hermaphroditism, origin of, Mr. Darwin's theory of, and observations on, 339
Heterogamy, explained, 198; and sexuality, 243
Heteromorphic flowers explained, 203
Heterostylism, explained, 203; and diclinism, 228; and diœcism, 218; and degrees of fertility, 204, seqq.; origin of, 213; and sexuality, 244; structure of stigmas in, 216; unstable, in stamens of *Narcissus cernuus*, (fig. 37) 121
Homogamy, explained, 198; and anemophily, 269; fluctuating conditions about, 201
Homology, of appendages and axis,

INDEX. 345

309; explained, 285; origin of, 300
Homomorphic conditions, 203
Homostyled, flowers, explained, 203; forms of *Auricula*, 208; of *Primula Sinensis*, 209
Hooks of *Uncaria*, (fig. 46) 156
Hypertrophy, in animal kingdom, 88; cause of, 51, 88; effects of, in unions, 86, 87; form, a cause of, 105, seqq.; 116, seqq.; in Orchids, 87; of placentas, 307

I

Illegitimate, or homomorphic unions 206
Impatiens, secretive stipules of, (fig. 43) 140
Impregnation, a form of nutrition, 250
Inconspicuous flowers, 251, seqq.; anemophilous, 265; due to degeneracy, 251, seqq.; origin of, 282; self-fertilising, 253, seqq.
Insects, origin of species by agency of, 329; relative proportion of, in regular and irregular flowers, 102, 103, 314; visitors to *Compositæ*, 315
Irregularity, origin of, 103
Irritability. *See* Ant-plants, Appendages, Form, Protoplasm, Zygomorphism.

L

Laws, of alternation, 41; of colour, 174; of superposition, 41
Leaf, cabbage, excrescences on, 307; of *Coniferæ*, adnate and free, 84, 85; opposite and verticillate, 9; transition from opposite to verticillate, (fig. 2) 11, 17, 18. *See* Phyllotaxis.
Leaf-traces, of *Arabis albida*, (fig. 7) 39; compared with floral, 40

Liber-fibre, origin of, 250
Light, and cell-division, 154; influences of, on leaves, 154; on roots of Ivy, 155; on nucleus, 154; and sleep of calyx and corolla, 155
Lysimachia, anatomy of floral receptacle of, (fig. 19) 77

M

Mechanical forces, action on boughs, 125; on corolla, 126; on pear growth, 124; on stamens, 81, 82, 126; tissues, formation of, by, 155, seqq. *See* Irritability.
Metamorphosis, of bracts, 286; of calyx, 288; of corolla, 292, 302; of flowers, 285, 295; of pistil, 295; of stamens, 292, 298; of tentacles of *Drosera*, 307
Movements, in corolla, 160; of filaments, 159, 161, 162; of pistil, 162; of stamens, 162; of staminode, 161; of stigmas and styles, 159, 162

N

Narcissus cernuus, unstable heterostylism of, 121
Natural selection, difficulties of, 333; forms of, 330, seqq.; insufficient as a cause, 335. *See* Selection.
Nectaries, 140, seqq.; and adelphous stamens, 58; irritation, an origin of, 141, 143; and pollination, 148; position of, 140, seqq.; staminal in *Atragene*, (fig. 44) 141; stipular in *Impatiens*, (fig. 43) 140
Nepenthes, origin of pitcher of, 146
Nucleus, effect of electrical irritation on, 154, of light on, 154; of pollen-tube, effect of, 250
Numbers, illustrations of special, 25–38; origin of, 9; principle of 4, 7

O

Obdiplostemony, 188; cause of, 190; origin of, 150
Opposite and verticillate leaves, 9; as origin of alternate, 11
Orchids, adhesive roots of, (fig. 42) 137; conducting tissue of, 165; degeneracy in, 172, 280, 319; effect of irritations on, mechanical, 114, of larvæ, 171, physiological, of pollen-tubes, 165, seqq.; hypertrophy in, 87; monstrous, 87; self-fertilising, 253, 318
Order of development of floral whorls, relative only, 195
Organs, floral, slow development of, 122; rudimentary, 283
Origin of species, fertilisation and, 329; by natural selection, 333; by response of protoplasm to environment, 3, 50, 51, 84, seqq., 88, 103, seqq., 112, seqq., 116, seqq., 126, 133, seqq.
Ovary, arrest of, 169; growth from irritation of larvæ, 171, from mechanical irritations, 114; from pollen-tube, 170, seqq.
Ovules, basilar, interpretation of, 74; of *Beta* (fig. 16), 73; emergence of, 195; homology of, 303, seqq.; foliaceous, (fig. 85) 305; metamorphoses of, 305, seqq.; of Orchids, 166, seqq., 281; order of development, interpretation of, 196; origin of, 303, seqq.; phyllody of, 302, (fig. 85) 305, (fig. 86) 306

P

Pansy, stigma and style of, (fig. 54) 255; self-fertilising forms, (fig. 55) 257
Pathfinders and colours, 178
Pear, cause of obliquity at base of, (fig. 38) 124; interpretation of receptacular tube of, 86, (fig. 22, *a*) 90, (fig. 26) 94; effect of tension and weight of, upon form of, 124
Pedicel, origin from peduncle in *Erodium*, 309
Pelargonium, anatomy of floral receptacle, (fig. 13) 65–67
Peloria, 128, seqq.; causes of, 130; and generic characters, 132; hereditary, 131; and hypertrophy, 131; induced by *Tingidæ*, 130
Perianth, excrescence on, (fig. 87) 306; form of, 101
Perigynous condition, 78
Petals, adhesion of, 78, seqq.; cohesion of, 56, seqq.; colours of, 174, seqq.; 253, 270, (*see* Colours); irritability of, 158, (fig. 47) 160, (fig. 49) 162. *See* Corolla.
Phyllody, of carpels, 302; of floral whorls, 301, seqq.; of ovules, 302. *See* Ovule.
Phyllotaxis, and æstivation, (fig. 3) 15; and arrangement, 39, seqq.; and number, 9, seqq.; and origin of flowers, 339
Phyteuma, cohesions of, (fig. 9) 50
Pistil, carpels, number of, 4, 7, seqq.; superposition of, 46, 47, in *Campanulaceæ*, 44; degeneracy in, 278; development of, rate of, 192; fibro-vascular cords of, (figs. 12, 13, 14, 15, 16) 64, 65, 68, 71, 73; metamorphoses of, 295, seqq.; movements of, 162; rationale of superposition in, 46, 47; syncarpous, 62, seqq. *See* Carpel, Gynœcium, Ovary, Receptacular Tube.
Pitcher of *Nepenthes*, origin of, 146, 307
Placenta, axile, 62; as a carpophore, 72; cords of, 64–77; free-central, 72, 76, (fig. 19) 77; hypertrophy of, 307; parietal, of Orchids, a sign of degeneracy, 281
Pollen, of anemophilous flowers, 267; of cleistogamous flowers, 258, seqq.; degeneracy of, 273, 276; of Orchids, 173; in ovules, 296; quan-

tity, reduction of, 273; of self-fertilising plants, 254
Pollen-tube, effects of, 166, 167; irritation due to, 164, seqq.; in Orchids, 166, seqq.; in *Oxalis*, 260; in *Verbascum*, 168; in Violets, 258; in Willows, 170
Pollination and nectaries, correlation between, 148
Polygamous flowers and environment, 242
Pressure, effects of mechanical, 101, seqq., 116, seqq., 123, seqq., 156, seqq.; resistance to, by cell-wall, 127
Primine and secundine, foliacious, 306
Primulaceæ, free-central placenta of, interpretation of, (figs. 18, 19) 76, 77
Principles, general, 1; of variation, 4
Protandry, cause of, 198; explained, 198; illustrations of, 191, seqq.; in *Echium*, (fig. 20) 82; and self-fertilisation, 272, 273, seqq.
Protogyny, anemophily as a cause of, 200, 269; causes of, 199, seqq.; emergence and order of development of flowers with, 195; explained, 198; inconspicuousness of many flowers with, 195
Protoplasm, common to animal and vegetable kingdoms, and phenomena same in both, 147; irritability of, to electricity, 152, to temperature, 153, to touch, 153, seqq.; origin of species due to responsive powers of (*see* Origin of Species); transmission of effects of irritation by continuity of, 163

R

Ranunculaceæ, arrangement in, illustrations of, 21; symmetry in, illustrations of, 21
Receptacle, floral, anatomy of, in *Cruciferæ*, (fig. 6) 32; in Hellebore, (fig. 12) 64; in Ivy, (fig. 14) 68; in *Lysimachia*, (fig. 19) 77; in *Pelargonium*, (fig. 13) 65; in *Primula*, (fig. 19) 77
Receptacular tube, 89, seqq.; anatomy of, in *Alstrœmeria*, (fig. 30) 97; arrested conditions in, 91, 100; with calyx foliaceous, (fig. 67) 289; of Cherry, (fig. 29) 97; of *Cotoneaster*, (fig. 22, *b*) 90; of *Fuchsia*, (fig. 27) 94; of *Galanthus*, 98; of Hawthorn, (fig. 25) 93; interpretation of, 86; morphological investigations of, 90; of *Mussænda*, (fig. 68) 290; of *Narcissus*, 98; of Orchids, (fig. 23) 92; of Pear, 86, (fig. 22, *a*) 90, (fig. 26) 94; of *Prunus*, (fig. 28) 95; of Rose, (fig. 24) 93; teratological investigations of, 92; views of, 89
Regularity, acquired, 128; explained, 5; observations on, 101; position of flowers with, 101; *Tingidæ* as causing, 130. *See* Peloria.
Resupination, origin of, 107
Roots, adhesive, of Orchids, (fig. 42) 137; origin of hairs on, 137; of Ivy, effects of light on, 155
Rudimentary organs, 283

S

Salvia, cleistogamous species, 262, 263; cords of sepals of, 55; filaments of, 268; self-fertilising species, 261
Scent, absence of, in self-fertilising flowers, 254
Secretive tissues, as conducting, 164, seqq.; irritation as a cause of, 142; of milk, 147; as nectaries, 140, seqq.; in *Nepenthes*, 146; origin of, 141
Secundine, and primine, foliaceous, 306
Seeds, character of, for double flowers, 299; number of, compared with carpels, 21, 278, with stamens, 275; proportion of, to seedlings in Orchids, 280

348 INDEX.

Selection, constitutional, 330, 334; experiment in, 331; by insects, 335; of the luckiest, 331; natural, 333. *See* Natural Selection.

Self-fertilisation, and the flora of Dovrefjeld, 259 : cosmopolitan, 283 ; Mr. Darwin's views on, 215, and review of, 315, seqq.; and degeneracy, 252 ; of *Epilobium*, (fig. 53) 255; general, 192, 199, 216 ; and homomorphism, 214; illustrations of, (figs. 52-60) 255-262 ; injuriousness of, disproved, 315, seqq.; misinterpretations regarding, 312, seqq.; of Orchids, 253, 318; peculiarities of, 253; rapid recovery of, 320; of *Stellaria media*, (fig. 52) 255; and whole colouring, 183

Sensitiveness, 151. *See* Protoplasm.

Sepaline cords, source of staminal and carpellary, 42, seqq.; in *Campanula*, (fig. 8) 43, and (fig. 15) 71; *Labiatæ* increase of, in calyx of, 56 ; *Salvia*, in calyx of, 55

Sepals, arrest of, 8 ; carpellary lobes of, in Pea, (fig. 70) 292; cords of, in *Campanula*, (fig. 8) 43, (fig. 15) 71, and in Hollyhock, 44 ; development of, order of, 191 seqq.; emergence of, 184, seqq., and in *Cruciferæ*, 32 ; foliaceous, in *Ranunculus* (fig. 66), in *Trifolium* (fig. 67), 289 ; homologous with petioles, 288; lateral pair of, in Crucifers, first to emerge, (fig. 6) 32, 185; nectaries superposed to, in Hellebore, (fig. 5) 22 ; numbers of, in whorls, 25, seqq.; ovuliferous, in Violet, (fig. 71) 292 ; petaloid, one abnormally in *Linaria*, (fig. 69) 291, normally in *Mussænda*, (fig. 68) 290 ; petals superposed to, in *Garidella*, (fig. 4) 21 ; pistiloid, 291 ; staminoid, 291 ; venation of, 289

Septa, absorption of, in liber and wood-fibres, 250 ; formation of, in pistils, 70, seqq.

Sex, sudden appearance of, 338; arrest of, 246; change of, in *Calendula*, 241 ; origin of, 246, 249 ; of seeds, 247; and soil, 239 ; and temperature, 237

Sexuality, in *Calendula*, 241 ; in *Centaurea*, 240 ; and environment, 230, 245 ; and heterogamy, 243 ; and heterostylism, 244 ; and nutrition, 233, seqq.; and soil, 239

Solution, explained, 5

Spring, in corolla of *Genista*, (fig. 47) 160 ; of stamens in *Medicago*, (fig. 49) 162; of styles 125, of *Viola*, (fig. 54) 255

Stamens, adelphous, and nectaries, 58 ; adhesion of, and mechanical forces, 81 ; cohesion of, 57 ; declinate, 110, 125, in *Dictamnus*, (fig. 33) 110 ; in *Echium*, (fig. 20), 82 ; in *Epilobium*, (fig. 34), 111 ; dimorphic, (fig. 37) 121 ; distribution of forces in, 81, 126 ; with heterostylism, 203, seqq.; irregularity in, origin of, 109; irritability of, 159, 161; movement of, 162; metamorphoses of, 292, 298; petaline, cause of absence of, 7, 20 ; whorls, number of, 8

Staminode, movement of, in *Lopezia*, (fig. 48) 161

Stigmas, of anemophilous flowers, 269 ; of *Aristolochia*, (fig. 21) 83 ; of heterostyled flowers, 216 ; irritability of, 115, 163 ; long-lived, 269 ; movements of, 162 ; by protoplasmic continuity, 163

Stimulus, produced by crossing, advantages of, 330 ; temporary effect of, 312, 330

Stipules, of *Acacia sphærocephala*, due to irritation, 157 ; nectariferous, of *Impatiens*, (fig. 43) 140

Strains, effect on boughs, (fig. 39) 125 ; and cohesions, 51, 53 ; hypertrophy by, in pears, (fig. 38) 124, in pedicels, 123, on stems, 123, on structures, 123 seqq.

Struggle for existence in seedlings, period of greatest, 330

Styles, hairs within, origin of, 139;
of heterostyled plants, 203, seqq.;
movement of springs in, of Pansy,
(fig. 54) 255; piston-action of,
(fig. 11) 60; of self-fertilising
plants, 254
Stylopod, placental origin of, 72
Superposition, of carpels, 44; laws
of, 41
Supportive tissues, 127
Symmetry, floral, changes in, 186;
decrease and increase of, 18; illustrations of, in *Ranunculaceæ*, 21;
and phyllotaxis, 14; variations of,
12
Syncarpous pistil, 62
Syngenesious anthers, 59

T

Tendrils, of *Ampelopsis*, 145; of
Cucurbitaceæ, 145; thickening of,
due to irritation, 156
Teratology, 2, 285, seqq.; 295, seqq.;
301, seqq.
Teucrium, structure of flower in adaptation to insects, 56, (fig. 36) 117
Trichomes, origin of, 133, seqq.
Trimorphic flowers, 210, seqq.
Typical flower, diagram and structure
of, (fig. 1) 3

U

Uncaria, hook of, (fig. 46) 156
Unions, cause of, 84; effect of hypertrophy in, 86; illegitimate, 206;
legitimate, 204
Unsymmetrical, corolla, 5; decrease
in floral whorls, 20

V

Variation, principles of, in flowers, 4
Vascular cords, in *Campanula*, (fig. 8)
43; as floral units, 300, 308, 309;
in *Malvaceæ*, 43; origin of, 42. See
Cords.
Versatile anthers, cause of, 268; in
wind-fertilised flowers, 266, seqq.
Verticillate and opposite leaves, 91
Vessels and cells, constructed to resist
pressure, 127; as supportive, 127
Violet, cleistogamous, (fig. 56) 258;
style and stigma of, in self-fertilising
forms, (fig. 55) 257
Virescence, explained, 301

W

Weeds, and fertilisation, 281; self-fertilising, cosmopolitan, 283
White flowers, 180; effect of crossing
with, 180; and self-fertilisation,
182
Whorls, floral, alternation of, 39, seqq.;
arrangement of, 39, seqq.; examples
of one to twelve membered, 25,
seqq.; illustrations from *Ranunculaceæ*, 21, seqq.; origin of, in
Cruciferæ, (fig. 6) 32; projected
cycles, 38; superposition of, 39,
seqq.; symmetrical increase and
decrease of, 18, and cause of, 19;
of typical flower, (fig. 1) 3
Wind-fertilised flowers. See Anemophilous and Anemophily.
Wood-fibre, origin of, 250

Z

Zygomorphism, origin of, 102, 116,
seqq.

32

D. APPLETON & CO.'S PUBLICATIONS.

ALEXANDER BAIN'S WORKS.

THE SENSES AND THE INTELLECT. By ALEXANDER BAIN. LL. D., Professor of Logic in the University of Aberdeen. 8vo. Cloth, $5.00.

The object of this treatise is to give a full and systematic account of two principal divisions of the science of mind—the senses and the intellect. The value of the third edition of the work is greatly enhanced by an account of the psychology of Aristotle, which has been contributed by Mr. Grote.

THE EMOTIONS AND THE WILL. By ALEXANDER BAIN, LL. D. 8vo. Cloth, $5.00.

The present publication is a sequel to the former one on "The Senses and the Intellect," and completes a systematic exposition of the human mind.

MENTAL SCIENCE. A Compendium of Psychology and the History of Philosophy. Designed as a Text-book for High-Schools and Colleges. By ALEXANDER BAIN, LL. D. 12mo. Cloth, leather back, $1.50.

The present volume is an abstract of two voluminous works, "The Senses and the Intellect" and "The Emotions and the Will," and presents in a compressed and lucid form the views which are there more extensively elaborated.

MORAL SCIENCE. A Compendium of Ethics. By ALEXANDER BAIN, LL. D. 12mo. Cloth, leather back, $1.50.

The present dissertation falls under two divisions. The first division, entitled The Theory of Ethics, gives an account of the questions or points brought into discussion, and handles at length the two of greatest prominence, the Ethical Standard and the Moral Faculty. The second division—on the Ethical Systems—is a full detail of all the systems, ancient and modern.

MIND AND BODY. Theories of their Relations. By ALEXANDER BAIN, LL. D. 12mo. Cloth, $1.50.

"A forcible statement of the connection between mind and body, studying their subtile interworkings by the light of the most recent physiological investigations."—*Christian Register.*

LOGIC, DEDUCTIVE AND INDUCTIVE. By ALEXANDER BAIN, LL. D. Revised edition. 12mo. Cloth, leather back, $2.00.

EDUCATION AS A SCIENCE. By ALEXANDER BAIN, LL. D. 12mo. Cloth, $1.75.

ENGLISH COMPOSITION AND RHETORIC. Enlarged edition. Part I. Intellectual Elements of Style. By ALEXANDER BAIN, LL. D., Emeritus Professor of Logic in the University of Aberdeen. 12mo. Cloth, leather back, $1.50.

ON TEACHING ENGLISH. With Detailed Examples and an Inquiry into the Definition of Poetry. By ALEXANDER BAIN, LL. D. 12mo. Cloth, $1.25.

PRACTICAL ESSAYS. By ALEXANDER BAIN, LL. D. 12mo. Cloth, $1.50.

New York: D. APPLETON & CO., 1, 3, & 5 Bond Street.

D. APPLETON & CO.'S PUBLICATIONS.

Professor JOSEPH LE CONTE'S WORKS.

EVOLUTION AND ITS RELATION TO RELIGIOUS THOUGHT. By JOSEPH LE CONTE, LL. D., Professor of Geology and Natural History in the University of California. With numerous Illustrations. 12mo. Cloth, $1.50.

"Much, very much has been written, especially on the nature and the evidences of evolution, but the literature is so voluminous, much of it so fragmentary, and most of it so technical, that even very intelligent persons have still very vague ideas on the subject. I have attempted to give (1) a very concise account of what we mean by evolution, (2) an outline of the evidences of its truth drawn from many different sources, and (3) its relation to fundamental religious beliefs." —*Extract from Preface.*

ELEMENTS OF GEOLOGY. A Text-book for Colleges and for the General Reader. By JOSEPH LE CONTE, LL. D. With upward of 900 Illustrations. New and enlarged edition. 8vo. Cloth, $4.00.

"Besides preparing a comprehensive text-book, suited to present demands, Professor Le Conte has given us a volume of great value as an exposition of the subject, thoroughly up to date. The examples and applications of the work are almost entirely derived from this country, so that it may be properly considered an American geology. We can commend this work without qualification to all who desire an intelligent acquaintance with geological science, as fresh, lucid, full, authentic, the result of devoted study and of long experience in teaching." —*Popular Science Monthly.*

RELIGION AND SCIENCE. A Series of Sunday Lectures on the Relation of Natural and Revealed Religion, or the Truths revealed in Nature and Scripture. By JOSEPH LE CONTE, LL. D. 12mo. Cloth, $1.50.

"We commend the book cordially to the regard of all who are interested in whatever pertains to the discussion of these grave questions, and especially to those who desire to examine closely the strong foundations on which the Christian faith is reared."—*Boston Journal.*

SIGHT: An Exposition of the Principles of Monocular and Binocular Vision. By JOSEPH LE CONTE, LL. D. With Illustrations. 12mo. Cloth, $1.50.

"Professor Le Conte has long been known as an original investigator in this department; all that he gives us is treated with a master-hand. It is pleasant to find an American book that can rank with the very best of foreign books on this subject."—*The Nation.*

COMPEND OF GEOLOGY. By JOSEPH LE CONTE, LL. D. 12mo. Cloth, $1.40.

New York: D. APPLETON & CO., 1, 3, & 5 Bond Street.

D. APPLETON & CO.'S PUBLICATIONS.

THE GEOLOGICAL HISTORY OF PLANTS. By Sir J. WILLIAM DAWSON, F. R. S. Vol. 61 of The International Scientific Series. With Illustrations. 12mo. Cloth, $1.75.

"The object of this work is to give, in a connected form, a summary of the development of the vegetable kingdom in geological time. To the geologist and botanist the subject is one of importance with reference to their special pursuits, and one on which it has not been easy to find any convenient manual of information."—*From the Preface.*

THE GEOGRAPHICAL AND GEOLOGICAL DISTRIBUTION OF ANIMALS. By ANGELO HEILPRIN, Professor of Invertebrate Paleontology at the Academy of Natural Sciences, Philadelphia, etc. Vol. 57 of The International Scientific Series. One vol., 12mo, 435 pages, $2.00.

"In the preparation of the following pages the author has had two objects in view: that of presenting to his readers such of the more significant facts connected with the past and present distribution of animal life as might lead to a proper conception of the relations of existing faunas; and, secondly, that of furnishing to the student a work of general reference, wherein the more salient features of the geography and geology of animal forms could be sought after and readily found."—*From the Preface.*

ANIMAL MAGNETISM. From the French of ALFRED BINET and CHARLES FÉRÉ. Vol. 59 of The International Scientific Series. 12mo. Cloth, $1.50.

"The authors, after giving a brief, clear, and instructive history of animal magnetism from its remotest known origin down through Mesmer and the Academic period to the present day, record their personal investigations among the hysterical, nervous, and generally supersensitive female patients in the great Paris hospital, La Salpêtrière, of which M. Féré is the assistant physician."—*Journal of Commerce.*

WEATHER: A POPULAR EXPOSITION OF THE NATURE OF WEATHER CHANGES FROM DAY TO DAY. By the Hon. RALPH ABERCROMBY, Fellow of the Royal Meteorological Society, London. Vol. 58 of The International Scientific Series. 12mo. Cloth, $1.75.

"Mr. Abercromby has for some years made the weather of Great Britain a special study, and has recently extended his experience by making a meteorological tour around the world. As a fruit of this preparation, he gives us a book that is to be commended for its simple, deliberate style, freedom from technicality and unnecessary theorizing, rational description, classification, and explanation of atmospheric phenomena, and rich store of illustration from the weather-maps of many parts of the world."—*The Nation.*

New York: D. APPLETON & CO., 1, 3, & 5 Bond Street.

CHARLES DARWIN'S WORKS.

ORIGIN OF SPECIES BY MEANS OF NATURAL SELECTION, OR THE PRESERVATION OF FAVORED RACES IN THE STRUGGLE FOR LIFE. Revised edition, with Additions. 12mo. Cloth, $2.00.

DESCENT OF MAN, AND SELECTION IN RELATION TO SEX. With many Illustrations. A new edition. 12mo. Cloth, $3.00.

JOURNAL OF RESEARCHES INTO THE NATURAL HISTORY AND GEOLOGY OF COUNTRIES VISITED DURING THE VOYAGE OF H. M. S. BEAGLE ROUND THE WORLD. New edition. 12mo. Cloth, $2.00.

EMOTIONAL EXPRESSIONS OF MAN AND THE LOWER ANIMALS. 12mo. Cloth, $3.50.

THE VARIATIONS OF ANIMALS AND PLANTS UNDER DOMESTICATION. With a Preface, by Professor Asa Gray. 2 vols. Illustrated. Cloth, $5.00.

INSECTIVOROUS PLANTS. 12mo. Cloth, $2.00.

MOVEMENTS AND HABITS OF CLIMBING PLANTS. With Illustrations. 12mo. Cloth, $1.25.

THE VARIOUS CONTRIVANCES BY WHICH ORCHIDS ARE FERTILIZED BY INSECTS. Revised edition, with Illustrations. 12mo. Cloth, $1.75.

THE EFFECTS OF CROSS AND SELF FERTILIZATION IN THE VEGETABLE KINGDOM. 12mo. Cloth, $2.00.

DIFFERENT FORMS OF FLOWERS ON PLANTS OF THE SAME SPECIES. With Illustrations. 12mo. Cloth, $1.50.

THE POWER OF MOVEMENT IN PLANTS. By Charles Darwin, LL. D., F. R. S., assisted by Francis Darwin. With Illustrations. 12mo. Cloth, $2.00.

THE FORMATION OF VEGETABLE MOULD THROUGH THE ACTION OF WORMS. With Observations on their Habits. With Illustrations. 12mo. Cloth, $1.50.

New York: D. APPLETON & CO., 1, 3, & 5 Bond Street.

D. APPLETON & CO.'S PUBLICATIONS.

THOMAS H. HUXLEY'S WORKS.

SCIENCE AND CULTURE, AND OTHER ESSAYS. 12mo. Cloth, $1.50.

THE CRAYFISH: AN INTRODUCTION TO THE STUDY OF ZOÖLOGY. With 82 Illustrations. 12mo. Cloth, $1.75.

SCIENCE PRIMERS: INTRODUCTORY. 18mo. Flexible cloth, 45 cents.

MAN'S PLACE IN NATURE. 12mo. Cloth, $1.25.

ON THE ORIGIN OF SPECIES. 12mo. Cloth, $1.00.

MORE CRITICISMS ON DARWIN, AND ADMINISTRATIVE NIHILISM. 12mo. Limp cloth, 50 cents.

MANUAL OF THE ANATOMY OF VERTEBRATED ANIMALS. Illustrated. 12mo. Cloth, $2.50.

MANUAL OF THE ANATOMY OF INVERTEBRATED ANIMALS. 12mo. Cloth, $2.50.

LAY SERMONS, ADDRESSES, AND REVIEWS. 12mo. Cloth, $1.75.

CRITIQUES AND ADDRESSES. 12mo. Cloth, $1.50.

AMERICAN ADDRESSES; WITH A LECTURE ON THE STUDY OF BIOLOGY. 12mo. Cloth, $1.25.

PHYSIOGRAPHY: AN INTRODUCTION TO THE STUDY OF NATURE. With Illustrations and Colored Plates. 12mo. Cloth, $2.50.

HUXLEY AND YOUMANS'S ELEMENTS OF PHYSIOLOGY AND HYGIENE. By T. H. HUXLEY and W. J. YOUMANS. 12mo. Cloth, $1.50.

New York: D. APPLETON & CO., 1, 3, & 5 Bond Street.

D. APPLETON & CO.'S PUBLICATIONS.

JOHN TYNDALL'S WORKS.

ESSAYS ON THE FLOATING MATTER OF THE AIR, in Relation to Putrefaction and Infection. 12mo. Cloth, $1.50.

ON FORMS OF WATER, in Clouds, Rivers, Ice, and Glaciers. With 35 Illustrations. 12mo. Cloth, $1.50.

HEAT AS A MODE OF MOTION. New edition. 12mo. Cloth, $2.50.

ON SOUND: A Course of Eight Lectures delivered at the Royal Institution of Great Britain. Illustrated. 12mo. New edition. Cloth, $2.00.

FRAGMENTS OF SCIENCE FOR UNSCIENTIFIC PEOPLE. 12mo. New revised and enlarged edition. Cloth, $2.50.

LIGHT AND ELECTRICITY. 12mo. Cloth, $1.25.

LESSONS IN ELECTRICITY, 1875-'76. 12mo. Cloth, $1.00

HOURS OF EXERCISE IN THE ALPS. With Illustrations. 12mo. Cloth, $2.00.

FARADAY AS A DISCOVERER. A Memoir. 12mo. Cloth, $1.00.

CONTRIBUTIONS TO MOLECULAR PHYSICS in the Domain of Radiant Heat. $5.00.

SIX LECTURES ON LIGHT. Delivered in America in 1872-'73. With an Appendix and numerous Illustrations. Cloth, $1.50.

FAREWELL BANQUET, given to Professor Tyndall, at Delmonico's, New York, February 4, 1873. Paper, 50 cents.

ADDRESS delivered before the British Association, assembled at Belfast. Revised with Additions, by the author, since the Delivery 12mo. Paper, 50 cents.

New York: D. APPLETON & CO., 1, 3, & 5 Bond Street.

D. APPLETON & CO.'S PUBLICATIONS.

GEORGE J. ROMANES'S WORKS.

JELLY-FISH, STAR-FISH, AND SEA-URCHINS. Being a Research on Primitive Nervous Systems. 12mo. Cloth, $1.75.

"Although I have throughout kept in view the requirements of a general reader, I have also sought to render the book of service to the working physiologist, by bringing together in one consecutive account all the more important observations and results which have been yielded by this research."—*Extract from Preface.*

"A profound research into the laws of primitive nervous systems conducted by one of the ablest English investigators. Mr. Romanes set up a tent on the beach and examined his beautiful pets for six summers in succession. Such patient and loving work has borne its fruits in a monograph which leaves nothing to be said about jelly-fish, star-fish, and sea-urchins. Every one who has studied the lowest forms of life on the sea-shore admires these objects. But few have any idea of the exquisite delicacy of their structure and their nice adaptation to their place in nature. Mr. Romanes brings out the subtile beauties of the rudimentary organisms, and shows the resemblances they bear to the higher types of creation. His explanations are made more clear by a large number of illustrations. While the book is well adapted for popular reading it is of special value to working physiologists."—*New York Journal of Commerce.*

"A most admirable treatise on primitive nervous systems. The subject-matter is full of original investigations and experiments upon the animals mentioned as types of the lowest nervous developments."—*Boston Commercial Bulletin.*

"Mr. George J. Romanes has already established a reputation as an exact and comprehensive naturalist, which his later work, 'Jelly-Fish, Star-Fish, and Sea-Urchins,' fully confirms. These marine animals are well known upon our coasts, and always interest the on-lookers. In this volume (one of the 'International Scientific Series') we have the whole story of their formation, existence, nervous system, etc., made most interesting by the simple and non-professional manner of treating the subject. Illustrations aid the text, and the professional student, the naturalist, all lovers of the rocks, woods, and shore, as well as the general reader, will find instruction as well as delight in the narrative."—*Boston Commonwealth.*

ANIMAL INTELLIGENCE. 12mo. Cloth, $1.75.

"A collection of facts which, though it may merely amuse the unscientific reader, will be a real boon to the student of comparative psychology, for this is the first attempt to present systematically the well-assured results of observation on the mental life of animals."—*Saturday Review.*

MENTAL EVOLUTION IN ANIMALS. With a Posthumous Essay on Instinct, by CHARLES DARWIN. 12mo. Cloth, $2.00.

"Mr. Romanes has followed up his careful enumeration of the facts of 'Animal Intelligence,' contributed to the 'International Scientific Series,' with a work dealing with the successive stages at which the various mental phenomena appear in the scale of life. The present installment displays the same evidence of industry in collecting facts and caution in co-ordinating them by theory as the former."—*The Athenæum.*

New York: D. APPLETON & CO., 1, 3, & 5 Bond Street.

D. APPLETON & CO.'S PUBLICATIONS.

DR. HENRY MAUDSLEY'S WORKS.

BODY AND WILL: Being an Essay concerning Will in its Metaphysical, Physiological, and Pathological Aspects. 12mo. Cloth, $2.50.

BODY AND MIND: An Inquiry into their Connection and Mutual Influence, specially in reference to Mental Disorders. 1 vol., 12mo. Cloth, $1.50.

PHYSIOLOGY AND PATHOLOGY OF MIND:

PHYSIOLOGY OF THE MIND. New edition. 1 vol., 12mo. Cloth, $2.00. CONTENTS: Chapter I. On the Method of the Study of the Mind.—II. The Mind and the Nervous System.—III. The Spinal Cord, or Tertiary Nervous Centres; or, Nervous Centres of Reflex Action.—IV. Secondary Nervous Centres; or, Sensory Ganglia; Sensorium Commune.—V. Hemispherical Ganglia; Cortical Cells of the Cerebral Hemispheres; Ideational Nervous Centres; Primary Nervous Centres; Intellectorium Commune.—VI. The Emotions.—VII. Volition.—VIII. Motor Nervous Centres, or Motorium Cummune and Actuation or Effection.—IX. Memory and Imagination.

PATHOLOGY OF THE MIND. Being the Third Edition of the Second Part of the "Physiology and Pathology of Mind," recast, enlarged, and rewritten. 1 vol., 12mo. Cloth, $2.00. CONTENTS: Chapter I. Sleep and Dreaming.—II. Hypnotism, Somnambulism, and Allied States.—III. The Causation and Prevention of Insanity: (A) Etiological.—IV. The same continued.—V. The Causation and Prevention of Insanity: (B) Pathological.—VI. The Insanity of Early Life.—VII. The Symptomatology of Insanity.—VIII. The same continued.—IX. Clinical Groups of Mental Disease.—X. The Morbid Anatomy of Mental Derangement.—XI. The Treatment of Mental Disorders.

RESPONSIBILITY IN MENTAL DISEASE. (International Scientific Series.) 1 vol., 12mo. Cloth, $1.50.

"The author is at home in his subject, and presents his views in an almost singularly clear and satisfactory manner. . . . The volume is a valuable contribution to one of the most difficult and at the same time one of the most important subjects of investigation at the present day."—*New York Observer.*

"Handles the important topic with masterly power, and its suggestions are practical and of great value."—*Providence Press.*

New York: D. APPLETON & CO., 1, 3, & 5 Bond Street.

D. APPLETON & CO.'S PUBLICATIONS.

SIR JOHN LUBBOCK'S (Bart.) WORKS.

THE ORIGIN OF CIVILIZATION AND THE PRIMITIVE CONDITION OF MAN, MENTAL AND SOCIAL CONDITION OF SAVAGES. Fourth edition, with numerous Additions. With Illustrations. 8vo. Cloth, $5.00.

"The first edition of this work was published in the year 1870. The work has been twice revised for the press in the interval, and now appears in its fourth edition *enlarged to the extent of nearly two hundred pages*, including a full index."

"This interesting work—for it is intensely so in its aim, scope, and the ability of its author—treats of what the scientists denominate *anthropology*, or the natural history of the human species ; the complete science of man, body and soul, including sex, temperament, race, civilization, etc."—*Providence Press.*

PREHISTORIC TIMES, AS ILLUSTRATED BY ANCIENT REMAINS AND THE MANNERS AND CUSTOMS OF MODERN SAVAGES. Illustrated. Entirely new revised edition. 8vo. Cloth, $5.00.

The book ranks among the noblest works of the interesting and important class to which it belongs. As a *résumé* of our present knowledge of prehistoric man, it leaves nothing to be desired. It is not only a good book of reference but the best on the subject.

"This is, perhaps, the best summary of evidence now in our possession concerning the general character of prehistoric times. The Bronze Age, The Stone Age, The Tumuli, The Lake Inhabitants of Switzerland, The Shell Mounds, The Cave Man, and The Antiquity of Man, are the titles of the most important chapters."—*Dr. C. K. Adams's Manual of Historical Literature.*

ANTS, BEES, AND WASPS. A Record of Observations on the Habits of the Social Hymenoptera. With Colored Plates. 12mo. Cloth, $2.00.

"This volume contains the record of various experiments made with ants, bees, and wasps during the last ten years, with a view to test their mental condition and powers of sense. The principal point in which Sir John's mode of experiment differs from those of Huber, Forel, McCook, and others, is that he has carefully watched and marked particular insects, and has had their nests under observation for long periods—one of his ants' nests having been under constant inspection ever since 1874. His observations are made principally upon ants, because they show more power and flexibility of mind; and the value of his studies is that they belong to the department of original research."

"We have no hesitation in saying that the author has presented us with the most valuable series of observations on a special subject that has ever been produced, charmingly written, full of logical deductions, and, when we consider his multitudinous engagements, a remarkable illustration of economy of time. As a contribution to insect psychology, it will be long before this book finds a parallel."—*London Athenæum.*

New York: D. APPLETON & CO., 1, 3, & 5 Bond Street.

D. APPLETON & CO.'S PUBLICATIONS.

APPLETONS' PHYSICAL GEOGRAPHY. Illustrated with engravings, diagrams and maps in color, and including a separate chapter on the geological history and the physical features of the United States. By JOHN D. QUACKENBOS, A. M., M. D., Adjunct Professor of the English Language and Literature, Columbia College, New York, *Literary Editor;* JOHN S. NEWBERRY, M. D., LL. D., Professor of Geology and Paleontology, Columbia College; CHARLES H. HITCHCOCK, Ph. D., Professor of Geology and Mineralogy, Dartmouth College; W. LE CONTE STEVENS, Ph. D., Professor of Physics, Packer Collegiate Institute; HENRY GANNETT, E. M., Chief Geographer of the United States Geological Survey; WILLIAM H. DALL, of the United States National Museum; C. HART MERRIAM, M. D., Ornithologist of the Department of Agriculture, NATHANIEL L. BRITTON, E. M., Ph. D., Lecturer in Botany, Columbia College; GEORGE F. KUNZ, Gem Expert and Mineralogist with Messrs. Tiffany & Co., New York; Lieutenant GEORGE M. STONEY, Naval Department, Washington. Large 4to. Cloth, $1.90.

APPLETONS' ATLAS OF THE UNITED STATES. Consisting of General Maps of the United States and Territories, and a County Map of each of the States, all printed in Colors, together with Railway Maps and Descriptive Text Outlining the History, Geography, and Political and Educational Organization of the States, with latest Statistics of their Resources and Industries. Imperial 8vo, cloth. $1.50.

THE EARTH AND ITS INHABITANTS. By ELISÉE RECLUS. Translated and edited by E. G. Ravenstein. With numerous Illustrations, Maps, and Charts.

M. Reclus the distinguished French Geographer has given in this work the most thorough and comprehensive treatise on the countries of the world yet produced. Maps, plans, and illustrations are lavish. It is subdivided as follows:

EUROPE, in 5 volumes. Imperial 8vo.
ASIA, in 4 volumes. Imperial 8vo.
AFRICA, in 3 volumes. Imperial 8vo.
AMERICA. (*In preparation.*)

Price, $6.00 per volume in library binding. Sold only by subscription.

A NEW PHYSICAL GEOGRAPHY. By ELISÉE RECLUS. In two volumes. Vol. I. The Earth. Vol. II. The Ocean, Atmosphere, and Life. With Maps and Illustrations. Price, $6.00 per volume, library binding. Sold only by subscription.

New York: D. APPLETON & CO., 1, 3, & 5 Bond Street.

www.ingramcontent.com/pod-product-compliance
Lightning Source LLC
Chambersburg PA
CBHW030407230426
43664CB00007BB/777